ATMOSPHERIC

THE BURNING STORY OF **CLIMATE CHANGE**

CAROLE WILKINSON

black dog books

To the members of Yarra Climate Action Now,
past and present – ordinary people concerned
about the state of our atmosphere and
how we leave it for future generations …
and willing to do something about it.

ATMOSPHERIC

Through greed, we have established an economy that destroys the web of life. We have changed our climate and drown in despair. Let oceans of justice flow. May we learn to sustain and renew the life of our Mother Earth.

Archbishop Desmond Tutu,
prayer for the People's Climate March,
September 2014

As with all the books in the Drum series, each chapter of *Atmospheric* begins with a short fictional piece. This is to help put young readers in the shoes of those who played various roles or witnessed different periods in the story of climate change. These pieces are all inspired by real events and many of the characters are real people, though some of the supporting characters are fictional.

All information throughout this book is as accurate as possible at time of publication.

First published in 2015 by **black dog books**,
an imprint of Walker Books Australia Pty Ltd
Locked Bag 22, Newtown
NSW 2042 Australia
www.walkerbooks.com.au

The moral rights of the author have been asserted.

National Library of Australia Cataloguing-in-Publication entry:
Wilkinson, Carole, 1950– author.
Atmospheric / Carole Wilkinson.
ISBN: 978 1 925126 37 2 (paperback)
Series: Drum.
For young adults.
Subjects: Climatic changes – Environmental aspects.
Weather.
Atmosphere.
551.6

Typeset in Adobe Garamond Pro and Agenda
Printed and bound in Australia by Griffin Press

The paper this book is printed on is certified against the Forest Stewardship Council® Standards. Griffin Press – a member of the Opus Group holds chain of custody certification SCS-COC-001185. FSC® promotes environmentally responsible, socially beneficial and economically viable management of the world's forests.

CONTENTS

INTRODUCTION

It's a crisp winter morning. The Sun is warming my face as I watch it slowly rise through a blue sky and a thin band of white clouds mottled with grey. What I'm looking at is part of our atmosphere.

The Earth's atmosphere is like a protective blanket holding in the Sun's warmth. Without it, the Earth would lose all the heat from the Sun and the Earth would be a very cold place. Our atmosphere provides exactly the right level of oxygen for us to breathe. It holds water that we need to live. It makes life on Earth possible.

We can't survive without it. And yet most of the time we don't even notice it. We treat our precious atmosphere as if it were a great big rubbish dump for the emissions we produce every day – when we use electricity and drive in our cars.

WHETHER THE WEATHER

The weather is the way our atmosphere interacts with the Earth's surface. Changes in atmospheric pressure cause the air above us to move around, creating winds and cyclones. The atmosphere contains water vapour that forms clouds. The water in clouds falls onto the Earth – sometimes as a gentle shower of rain, other times as flooding rain. When the atmosphere gets very cold, it falls as hail or snow.

Talking about the weather was once something we did when we couldn't think of anything else to say. It was considered boring conversation – small talk. Nowadays extreme weather – cyclones, floods, heatwaves – is big news. It makes world headlines. And more than ever, the weather has become life-threatening.

Over the past few decades, weather has become the most important topic in the world.

Weather is the atmospheric conditions that we experience in the short-term – today, or last week – in the place where we live. Climate is the pattern of weather in a larger region or across the entire planet, and over a longer period of time – the average weather over decades, centuries, even millennia.

WEATHER WORRIES

The atmosphere is warming. Scientific evidence shows that the average temperature on Earth is now hotter than at any time in human civilisation. The temperature rise is causing longer, hotter heatwaves; more intense cyclones; and more frequent floods – affecting people all over the world.

Climate scientists have proved that we, human beings, are responsible for the rising temperatures. Not everyone agrees with them. Some people continue to think there is no danger.

The weather has started a global debate. On one side are those who don't accept the science – a few scientists, business people in industries that rely on fossil fuels and people who are not convinced by the scientific arguments. On the other side are the majority of scientists who work in the field of climate science (97% of them), and those who trust them.

I'm one of the people greatly concerned about our climate and what we are doing to our atmosphere. That's why I have written this book.

1

A NARROW BAND OF BLUE

21 September 2014
Melbourne, Australia

Mum was driving me to the State Library. I had an appointment to play chess with Vasily, but when we got to the city, the streets were blocked by police barricades.

"Is there a bomb threat?" I asked.

Mum pointed to people with banners. "No, there's some sort of rally. You'll have to walk the rest of the way."

I jumped out of the car, but the crowds grew as I got closer to the library. There were people with signs and banners, someone in a polar bear suit. There was an enormous puppet of the prime minister looking like a gangster. About 50 people tried to get me to take flyers and sign petitions. I elbowed my way through the crowd towards the steps of the library. And wouldn't you know it – that's where the rally was being held.

I found Vasily leaning on one of the columns out the front of the library. The middle four columns had been covered with grey crepe paper with something fluffy at the top, so that they looked like chimneys at a power station. How did they even do that?

"What's going on?" I asked, squeezing in alongside Vasily.

"Climate change rally," he said.

Vasily is an old Russian guy. We play chess twice a week – not in a quiet room, but outdoors, using the giant chess set on the library forecourt. We like to put on a performance for the people who watch us. Vasily swears in Russian when I take one of his pieces. I do a few dance moves when I win (not very often).

"No chance of a game today with all these crazy people here," I said. I was annoyed.

Vasily shrugged his shoulders. "Come," he said. "I will buy you a drink."

We went to the cafe inside the library. Vasily ordered a thick black coffee. I got an energy drink.

His phone beeped. It was ancient, at least ten years old. He was trying to find the text message. I can't stand watching old people who don't know how to use technology.

"Let me do it." He handed me the phone. "Someone called Sofia is looking for you. Shall I tell her where you are?"

He nodded. About ten seconds later, someone flopped down next to us. It was a girl about my age. She looked nice. Vasily did the introductions. She's his granddaughter.

"Do you want something to eat?" he asked.

I was hungry. I scanned the menu. It was all avocado baguettes and zucchini frittata.

"Nah, I'll have a burger later. When I get my new runners."

Sofia looked down at my feet. "What's wrong with those?"

"They're … old." I thought it was pretty obvious. I'd had them for nearly two years.

She looked at my empty energy drink bottle. Next thing I know, she's giving me this lecture about "clothing miles" and waste and methane and Brazilian rainforests.

"It's not even proven that climate change exists," I said.

She looked at me like I'd spat in her herbal tea.

"You need to be better informed." She shoved a pamphlet in my face and then stomped off.

Vasily and I finished our drinks and went outside. The crowd was even bigger. I was looking around trying to work out the best way to make an escape, but the crowd was blocking all the streets around the library.

There was the squeal of microphone feedback and then cheering as the first speaker was introduced. It was Sofia, and she'd chained herself to one of the columns.

"Why'd she do that?" I asked Vasily.

"To get media attention," Vasily said, as if it was a perfectly reasonable thing to do.

Sofia started going on about how climate change will, you know, end the world, how everyone should be doing something. I don't know what. How the atmosphere is full of greenhouse gases. I looked up at the blue sky. It looked all right to me. Vasily was listening to her, nodding.

Sofia finished and people cheered. Vasily clapped. Someone else stepped up to give a speech. Sofia was still chained to the column. Three policemen walked over to her with a pair of boltcutters. A news crew was making its way through the crowd.

"Won't she get arrested?" I said.

"Yes."

I looked at the pamphlet. Obviously this was something she thought worth getting arrested for.

Vincent Dwyer, student, aged 15

LAYERS OF THE ATMOSPHERE

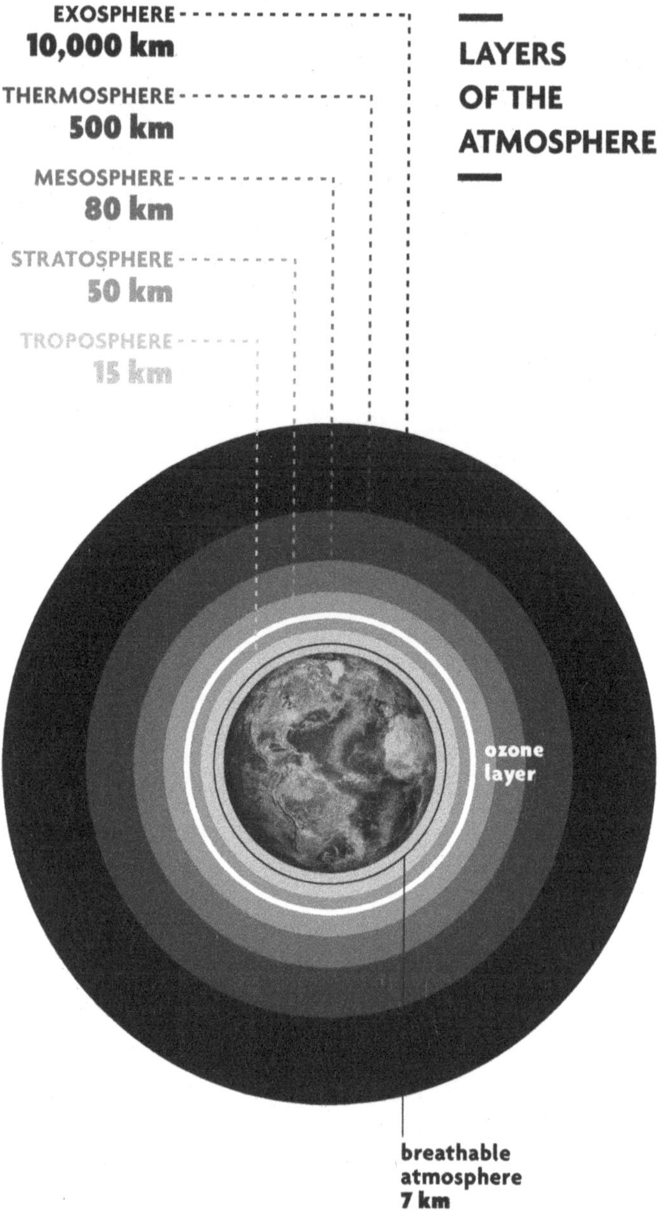

EXOSPHERE
10,000 km

THERMOSPHERE
500 km

MESOSPHERE
80 km

STRATOSPHERE
50 km

TROPOSPHERE
15 km

ozone layer

breathable atmosphere
7 km

Climate change is complex. The knowledge we have about what makes up our atmosphere, and how humans affect it, comes from hundreds of years of scientific study conducted by thousands of scientists from fields as diverse as geology, astronomy, chemistry, geophysics and paleontology.

Before we can understand climate change, we need to understand our atmosphere.

KEEPING US WARM

The Sun is Earth's heater. It bathes our planet in solar radiation (which we call sunlight). We experience it as light and warmth. Our atmosphere stops about half of the Sun's energy from reaching us, either by absorbing it or reflecting it back into space. Solar energy is made up of visible light (which enables us to see) and other types of light – ultraviolet and infra-red. The atmosphere allows most of the visible light to pass through.

Ultraviolet radiation is what gives us sunburn. The Earth's atmosphere absorbs most of that ultraviolet, but enough gets through that we still have to protect our skin from strong sunlight. A lot of the infra-red radiation is also absorbed by our atmosphere.

THE AIR WE BREATHE

The atmosphere is the layer of gases that extends up from the Earth's surface about 10,000 kilometres to the very edge of space. There are five zones in the atmosphere:

the troposphere, the stratosphere, the mesosphere, the thermosphere and the exosphere.

We live in the troposphere. This narrow band of the atmosphere is only 10 to 18 kilometres deep. Seventy-five per cent of the gases in the atmosphere are concentrated in the troposphere. The level of oxygen we require to breathe naturally is found only in the few kilometres in the bottom half of the troposphere.

Most of our atmosphere is made up of oxygen (21 per cent) and nitrogen (78 per cent). The other one per cent of the atmosphere consists of what are called "trace gases", because they are present in such small amounts. We call the mix of gases in the troposphere "air".

THE GREENHOUSE EFFECT

The solar energy that reaches the Earth is absorbed by the land, the oceans and the vegetation. It is transformed into heat (in the form of infra-red), which then radiates out from the Earth back towards space. If there was nothing to stop it, all of that cosy heat would dissipate out into space and the Earth would be a very cold place. Luckily, some of the gases in our atmosphere prevent the heat from escaping, trapping it in our atmosphere and keeping us warm.

This is known as the greenhouse effect. It is a natural phenomenon that keeps the temperature on Earth in a range that allows the creatures of the world to survive.

THE GREENHOUSE EFFECT

1. Solar energy (sunlight) reaches Earth's atmosphere. Some is reflected back into space – by the atmosphere and by ice and snow on the Earth's surface.

2. About half of the solar energy reaching Earth is absorbed by the land, oceans and vegetation. The Earth warms up.

3. Some of the Earth's heat radiates towards space.

4. But some heat from Earth is trapped by greenhouse gases.

5. Human activities – mainly burning fossil fuels – increases the amount of greenhouse gases released into the atmosphere.

6. This greater concentration of greenhouse gases traps more heat in the atmosphere and the temperature on Earth rises.

GLOBAL WARMING

The oxygen and nitrogen in the atmosphere allows solar energy to pass straight through. Among the trace gases are carbon dioxide and other gases that do the job of stopping the heat from escaping, even though they make up just a fraction of a per cent of the atmosphere. These gases are known as the greenhouse gases.

We need the greenhouse gases and the greenhouse effect to keep our planet warm. But emissions produced by human activities, such as burning coal, gas and oil, and land clearing, are increasing the amount of greenhouse gases in the atmosphere, making our climate hotter. This is called global warming.

It took scientists 300 years, from the 1400s to the 1700s, just to identify the key gases that make up the air we breathe. Oxygen wasn't discovered until 1774.

THE MAIN CULPRIT

Carbon dioxide (CO_2) is the largest component of the greenhouse gases, and yet it makes up just 0.04 per cent of the atmosphere. It is such a small amount that it is measured in parts per million (ppm). Carbon dioxide is released into the atmosphere by natural processes: when animals (including us) exhale, when dead plants and animals rot, during bushfires and when volcanoes erupt. There are other natural processes that take up some of this carbon dioxide, including growth in plants and absorption of carbon dioxide by oceans. These stores of carbon are known as carbon "sinks". They take carbon dioxide out of the atmosphere and help keep it at a safe level.

But carbon dioxide produced by human activities has increased over the last 200 years. The carbon sinks can't absorb it all. This has altered the natural balance of the atmosphere.

GOOD CO_2

If there was no carbon dioxide in the atmosphere we'd all freeze to death. But carbon dioxide has another crucial function.

The average temperature on Earth is 14.5°C. Without the greenhouse gases in our atmosphere, the average temperature of the Earth would be around 30°C colder (-15°C).

The creatures that inhabit the Earth, including humans, need oxygen to live. When we inhale, our lungs take in oxygen from the air. This reacts with glucose from the food we eat to produce the energy that makes all the cells in our body work properly. Carbon dioxide is a by-product of that process. When we exhale, we expel carbon dioxide. We would eventually use up most of the oxygen in the atmosphere and die if nature hadn't created a way to replace it.

We have a symbiotic relationship with plants. They are opposite to us. Plants need carbon dioxide to survive. They absorb it through their leaves and it reacts with the visible light from the Sun to create carbohydrate, which plants

Carbon dioxide also makes bread rise, soft drinks fizzy and is used to make mist in movies.

need to grow. This process is called photosynthesis. One of the waste products from this reaction is the very thing that humans need to survive – oxygen. This beautiful balance replenishes the oxygen in the atmosphere, and enables us to keep breathing.

Other planets aren't as comfortable as Earth. On Mars the atmosphere is much thinner, so it is much colder. The average temperature is -60°C. The atmosphere on Venus is very thick compared to Earth's. That makes Venus the hottest planet in the solar system. Its average temperature is 450°C, which is the maximum temperature of an oven.

OTHER GREENHOUSE GASES

Carbon dioxide isn't the only greenhouse gas. Other important greenhouse gases are methane (CH_4) and nitrous oxide (N_2O). Water vapour is also a greenhouse gas. These all have natural sources and man-made sources that add to global warming.

There is another group of greenhouse gases that didn't exist before humans started manufacturing them. They include chlorofluorocarbons (CFCs) and other gases containing chlorine and fluorine; carbon tetrachloride (CCl_4), used in dry-cleaning and refrigeration; and sulphur hexafluoride (SF_6), used in the electrical industry. These synthetic greenhouse gases have had a dramatic effect on our atmosphere.

There is so little methane in the atmosphere that no one noticed it was there until 1948.

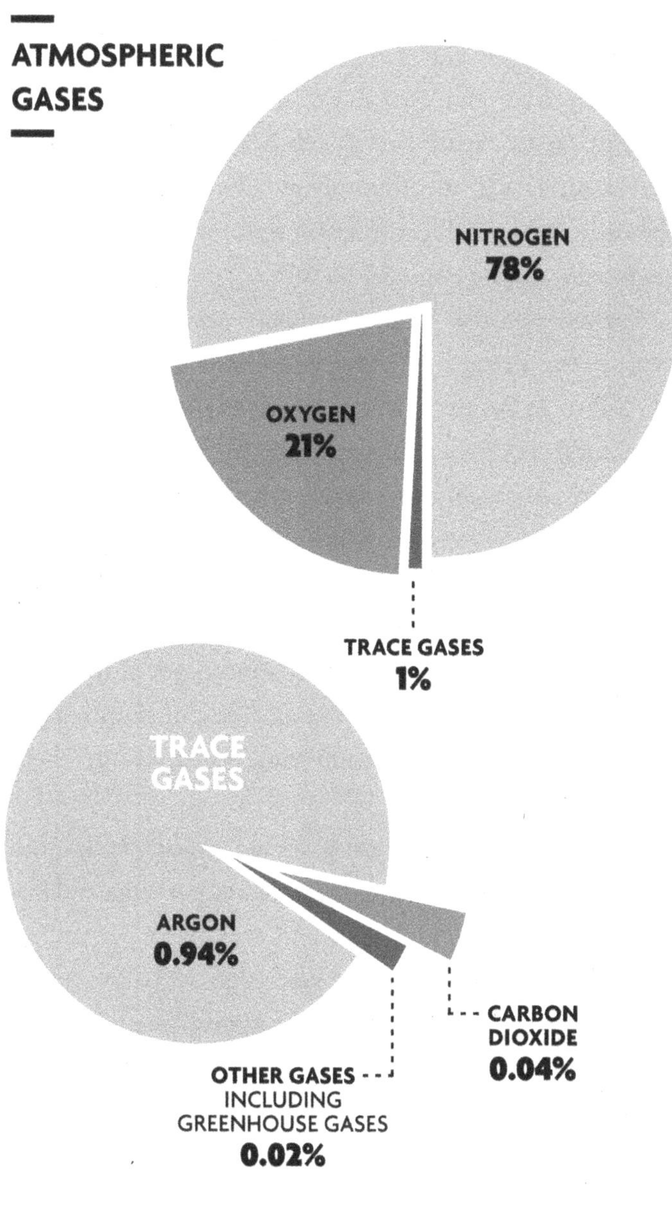
ATMOSPHERIC
GASES
NITROGEN
78%
OXYGEN
21%
TRACE GASES
1%
TRACE
GASES
ARGON
0.94%
CARBON
DIOXIDE
0.04%
OTHER GASES
INCLUDING
GREENHOUSE GASES
0.02%

GLOBAL WARMING POTENTIAL

Each greenhouse gas has a different ability to heat up the atmosphere and contribute to global warming. This depends on its capacity to absorb heat and the length of time it survives in the atmosphere before breaking down. The capacity for each gas to warm the atmosphere is called its global warming potential (GWP).

Carbon dioxide is the standard against which we measure the strength of other greenhouse gases, so it has a GWP of 1. Because carbon dioxide is such a key gas in the natural processes of life on Earth, some carbon dioxide emissions are quickly absorbed by the biosphere – by the oceans and by vegetation. Much of this carbon dioxide is only temporarily removed from the atmosphere, as other natural processes, such as animal respiration and bushfires, are replacing carbon dioxide in the atmosphere at the same rate. However, half of the carbon dioxide emissions remain in the atmosphere for about 300 years. A quarter stays there almost indefinitely.

The synthetic greenhouse gases are present in minute amounts in the atmosphere, measured in parts per million (ppm), parts per billion (ppb) and even parts per trillion (ppt). And yet some have a very high global warming potential because they survive in the atmosphere for tens of thousands of years.

WATER VAPOUR

Water is present on Earth in different forms – as a liquid in rivers, lakes and seas; as water droplets that make up clouds and fog; and as a solid in the form of ice at the poles and in glaciers. Water also exists in the air as a gas called water vapour, which forms when water evaporates.

Water vapour is the most abundant greenhouse gas in our atmosphere, but human activities directly add only a small amount.

However, humans are indirectly responsible for an increasing amount of water vapour in the atmosphere. The greenhouse gases that humans produce cause the atmosphere to warm. The warmer the air, the more water evaporates from the oceans, and this warmer air can hold more water vapour than cold air. When the atmosphere warms, the amount of water vapour in the atmosphere increases.

In hot, humid areas near the equator, the air can hold as much as four per cent water vapour, but in the cold polar regions it may only hold 0.2 per cent.

More water vapour in the air also means heavier rain, and can lead to an increase in severe flooding around the world and cyclones with more intense rain.

GREENHOUSE GASES IN EARTH'S ATMOSPHERE

GREENHOUSE GAS IN ATMOSPHERE	MAIN HUMAN SOURCES	CURRENT CONCENTRATION IN ATMOSPHERE	LIFE IN ATMOSPHERE IN YEARS	GLOBAL WARMING POTENTIAL
CARBON DIOXIDE (CO_2)	Burning fossil fuels, cement manufacture	401 parts per million	100-300	1
METHANE (CH_4)	Rice fields, food waste, leaks from gas manufacture and distribution, cattle	1768 parts per billion	12	28
NITROUS OXIDE (N_2O)	Fertilisers, vehicle emissions	326 parts per billion	121	265
SYNTHETIC GREENHOUSE GASES	Industrial processes, such as aluminium smelting, refrigeration, cleaning	4-500 parts per trillion	1-50,000	800-23,000

MAN-MADE CLIMATE CHANGE

The concern today about the changing climate is that it is changing faster – much faster than in the past. Climate has always changed, but climate scientists have discovered that for at least 800,000 years the average amount of carbon dioxide in the atmosphere always stayed between about 180 and 300 ppm. Until recently.

In the last 250 years, the amount of carbon dioxide in the atmosphere has increased rapidly. In May 2013, it reached 400 ppm for the first time in millions of years.

One of the last times carbon dioxide levels topped

400 ppm was during the geological period called the Miocene around 4 million years ago. During the middle of that period, 14–16 million years ago, temperatures were 3–6°C warmer, sea levels were between 25 and 40 metres higher than today, and modern humans were yet to evolve.

Climate scientists have estimated that 350 ppm is the maximum "safe" level for carbon dioxide in our atmosphere.

2

BURIED DEEP

1843

Wallsend Coalmine, Newcastle, England

The setting sun has turned the sky pink. I'm standing next to the winding tower at the pit head. The big wheels are turning, bringing the men of the day shift up. I soak up the last of the sunlight while I can.

The rest of the men and the young lads on the night shift gather near the shaft, each holding his water container and his bait. The other boys are larking about. My heart is thumping. I'm trying not to show I'm scared. The older men, the hewers, squat. After a life spent crouching in the pit, standing is hard on their stomach muscles. My Da is with them. He's got his back to me. I think he's worried I'll cry if he looks at me. He might be right. Today will be my first day working in the pit.

I collect a safety lamp from the lamp keeper. The cage rattles up and the men of the day shift come out. At this time of year, they never see the sun.

I get into the cage with the others, hoping they don't notice my legs shaking. I am going underground. Mr Dingley mutters a prayer. He survived the explosion of 1835. More than 100 men died that day. Some of the other men join in the Amen. I do too. The cage stops with a thump and we file out. The light from our lamps does nothing to chase away the darkness.

I am to be a putter. The deputy tells me what I must do. It sounds easy enough. Me and the other putter will haul the tub back and forth to the coalface. The men climb into the tub. I hitch myself to it with a harness. I feel like the darkness is crushing my body. The panic rises in my throat. The other putter is only a year or two older than me, but he acts like he's in charge. He thumps me in the chest.

"You better do your share," he says.

He pushes the tub and I pull it, into the depths of the Earth.

At the coalface, the hewers get out of the tub. It's so hot down here the men strip down until they're wearing just a pair of cut-off drawers, and I do the same. The tunnel is only high enough for a man to crouch in. The hewers work the seam with their picks. The other putter picks up the lumps of coal that the hewers hack out and puts them in the tub. I follow his lead. I can't stop thinking about all that earth and rock above my head, with just a few wooden props to stop it collapsing on me.

I sing a hymn. I whisper the words so the other boy won't hear me. My brothers laughed when Ma told me to sing if I got frightened. They both started work aged seven, employed as trappers who open doors in the tunnels to let the air flow through. They say I have it easy because I went to school and didn't have to start work until I turned ten.

When the tub is full, we haul it back to the shaft. It must be half a mile. And that's what I must do all day.

We stop for a break when the deputy tells us (we have no way of marking the time). And we have a drink from our water cans (it's too hot for tea) and I eat my bait – a slice of bread and jam. I sit by myself.

Then it's back to work hauling the tub. Each tub takes almost half a ton of coal, so I'm told. My shoulders hurt where the harness straps dig in. My hands are sore from picking up the lumps of coal. This is what I'll be doing for 12 hours a day, six days a week, until I'm 17 and I can become a hewer like Da.

My shift seems more like a week. On the last haul, the way the light from my lamp reflects off a small lump of coal catches my eye. I pick it up. In the dim light, I can see a faint pattern on one face of the lump. The other putter opens his mouth to yell at me to get back to work, but the whistle sounds the end of our shift. I put on my trousers and jacket and slip the lump of coal into my pocket. We haul the hewers back.

The cage rises up. The air cools, the darkness fades and then I'm back in the world at last. The rising sun is on the horizon. The light is blinding. All the men's faces are black as Africans. I look at my hands. They are black too. The white-faced day shift, including my two brothers, greets us with a nod, and I make my way home with Da. I'm so tired I can hardly put one foot in front of the other. Da walks alongside me with his hand on my shoulder. When the men are out of earshot, he says, "You did well, son."

Ma has a hot bath waiting for us in the scullery. Da goes first and the water is dark grey by the time it's my turn. I get in and the warmth soothes my sore body as I scrub off the grit and grime.

As Ma hangs up my jacket, she feels the lump of coal in my pocket. "What's this?" she says.

I'd forgotten about it. "Give it here," I say.

She hands it to me and I turn it over. There is the pattern I saw in the half-light. It's clear as anything in daylight. It looks like a fern leaf, delicate, beautiful and black. I don't know how it got there, but I'll keep it. A souvenir of my first day down the mine.

Billy Lowe, putter, aged ten

Much of the rise in carbon dioxide in the atmosphere is due to the fact that we burn fossil fuels, coal in particular. People started mining coal around 2000 years ago, but our awareness of the problems that burning coal causes is a relatively recent story in Earth's history. The story of the formation of fossil fuels, however, began long before humans existed.

FOSSIL FUELS

Coal, oil and gas are fossil fuels. They are hydrocarbons – compounds made up of carbon and hydrogen. About 80 per cent of the energy used in the world is made by burning fossil fuels. The use of fossil fuels by humans is the main cause of the global warming that is happening now.

Fossil fuels were formed beneath the surface of the Earth from the remains of dead plants and animals over a period of hundreds of millions of years.

Most of the electricity we use is generated in power stations that burn coal. At home we use electricity and gas for heating and cooking. Oil is refined to produce petrol for transport. Industry burns coal, oil and gas to manufacture things.

WHERE DOES THE WORLD'S ENERGY COME FROM?

FOSSIL FUELS 80.6%

RENEWABLE ENERGY 16.7%

NUCLEAR 2.7%

AN OLD STORY

There is nothing new about climate change. The climate has changed throughout Earth's long history. At times in the past, much more of the surface of the planet was covered in ice, so it was a lot colder than it is now. These periods lasted for many millions of years and are called ice ages. There have also been much longer periods when it was far warmer than today.

One warm period 300 million years ago has great significance for climate change today. It was called the Carboniferous period. Most of the Earth's coal was formed during this period.

CARBONIFEROUS

In the Carboniferous period (359–299 million years ago), not only was it warmer than it is today, but also the landmasses on Earth were in different places. Australia, along with Africa, India, Arabia, Antarctica and South America were joined in a huge landmass in the Southern Hemisphere, which we call Gondwana. In the Northern Hemisphere, what would eventually become Europe, North America and China were linked together and surrounded an ancient sea called the Tethys Ocean.

It was hundreds of millions of years before humans were living on Earth. There were no mammals at all, no birds, not even dinosaurs. The animal life in the Carboniferous period consisted of insects similar to those that exist today, such as millipedes, cockroaches and dragonflies. But they were huge. A millipede could be two metres long, spiders could span a metre. Giant dragonflies fluttered in the humid air. On land, there were amphibians that looked something like crocodiles. In the waters, there were fish and crustaceans.

An artist's idea of vegetation during the Carboniferous period.

"COOKING" COAL

Around the edges of the Tethys Ocean, the land was warm and humid with large areas of swamp. There were no trees or flowering plants at this time. Forests grew in the swamps, but they were not forests of trees; they were forests of simple plants, such as ferns, horsetails and club mosses. Some of these ancient plant species have survived to present times, although now they are usually small. In the Carboniferous period, they grew as tall as 30 metres.

> *"When we burn wood we release carbon that has been out of atmospheric circulation for a few decades, but when we burn fossil fuels we release carbon that has been out of circulation for eons."*
> **Tim Flannery, Founder of Climate Council, 2005**

When plants died and were covered with the stagnant waters of the swamps, the lack of oxygen prevented them from decomposing properly. More layers of soggy, dead plant material accumulated. Then they were buried by sediment – sand and clay washed into the swamps by rivers. Over thousands of years, the layer of dead plants turned into a moist, black substance called peat.

Millions of years passed and the sedimentary layer on top of the peat became deeper – and heavier, eventually turning into rock. The weight squeezed moisture out of the peat. Now deep underground, temperatures of over 100°C heated and compressed the peat until it gradually hardened and became the black rock that we call coal. Some scientists refer to this long process of coal formation as "cooking".

PLANKTON GRAVEYARD

Oil was formed from microscopic organisms that lived in oceans long ago. Crude oil formation began in different places around the world millions of years ago. Most of the world's oil was formed during the Triassic, Jurassic and Cretaceous periods (250–66 million years ago). These were also warmer periods of Earth's history. Dinosaurs were evolving on land. Pliosaurs and plesiosaurs swam in the oceans. But it was microscopic plants called phytoplankton living in oceans that were the ancestors of oil.

Oil in its natural state is called crude oil. It is a complex mixture of hydrocarbons.

At certain times during these periods the temperature rose even higher, melting the ice at the poles and warming the oceans. The amount of carbon dioxide in the air rose to about five times today's levels. There was more evaporation from the seas, which led to heavy rains and flooding. Large amounts of plant material were washed into the sea by swollen rivers. This acted like fertiliser and the result was an explosion of growth among the microscopic phytoplankton living in the oceans.

This upset the balance of marine life: there wasn't enough oxygen to sustain all that growth. Millions and millions of tonnes of phytoplankton died and sank to the bottom of the ocean. But in the stagnant oxygen-less depths, the dead phytoplankton couldn't decay. Instead, they collected on the ocean floor, forming a sticky black ooze.

Over millennia, bacteria converted the dead plankton into a substance called kerogen. Heat from the Earth and the weight of sedimentary layers turned the kerogen into black shale rock. The conditions in the Earth's crust had to be just right for oil to form – at a depth of two to four kilometres where the temperature was 50–100°C.

In America, crude oil is called petroleum and what Australians call petrol is called gasoline.

The immense weight of the sedimentary rocks squeezed the oil out of the shale. Crude oil is a thick black liquid, lighter than water. It moves. Not in streams, but in tiny amounts that find their way up through the porous rock above them, like water soaking into a sponge. Eventually, non-porous rocks called cap rocks trap the oil. Exploration companies drill down through the cap rocks to locate the oil.

DEEPER AND WARMER

Natural gas is 90 per cent methane, which is a colourless and odourless gas. It was formed in a similar way to oil, from the same organisms, often in the same locations during the same geological period. The conditions for gas formation were slightly different though. Gas formed deeper in the Earth, around three to six kilometres, where it is hotter, between 150 and 250°C. The gas also migrates upwards through porous rock. Because it is lighter, it collects above the oil.

Methane is not toxic, but it is very explosive. When oil was first discovered, the drillers burned off the gas found at the same sites because it was dangerous. Later, oil companies realised the commercial value of natural gas and began collecting it, transporting it to towns and cities, and selling it to households to use for lighting, heating and cooking.

UNDER DOWN UNDER

Australian fossil fuels have a different history. Our coal deposits were laid down about 50 million years after the Carboniferous period, in the Permian period (299–252 million years ago), which was much cooler. In Victoria, there are deposits of an even younger coal that originated in the Paleogene period, just 50–15 million years ago. This coal has not had so long to "cook". It is called brown coal and contains less carbon and more moisture than black coal. Burning brown coal produces only a quarter of the heat from burning black coal. It also produces 30 per cent more greenhouse gases.

The oil and gas deposits in Australia have different origins to deposits in other parts of the world. They weren't formed by minute plants in the ocean, but by the remains of land plants.

GEOLOGICAL TIMELINE FOR LAST 500 MILLION YEARS

EON	ERA	GEOLOGICAL PERIOD	EPOCH	WHEN THEY BEGAN (millions of years ago)	WHAT HAPPENED?
Phanerozoic	**Cenozoic**	Quaternary	Holocene	0.01	Human activities began affecting the atmosphere and climate.
			Pleistocene	2.6	Modern humans appeared. Last ice age began.
		Neogene	Pliocene	5.3	
			Miocene	23	Earliest humans appeared.
		Paleogene	Oligocene	34	
			Eocene	55.8	
			Paleocene	65.5	Dinosaurs died out. Oil formed in Australia.
	Mesozoic	Cretaceous		145	Oil formation. Primates appeared.
		Jurassic		200	Oil formation. First mammals and birds evolved.
		Triassic		251	Oil formation. Dinosaurs appeared.
	Paleozoic	Permian		299	Coal formation in Australia.
		Carboniferous		359	Coal formation in Northern Hemisphere. First reptiles evolved.
		Devonian		416	First land animals evolved.
		Silurian		444	
		Ordovician		488	Land plants evolved.
		Cambrian		542	First primitive animals with a backbone evolved.

3

SEEDS OF KNOWLEDGE

12,000 years ago
Mesopotamia

—

The young ones laugh and play, hiding from one another among the bushes. They stop suddenly, grab handfuls of berries and fill themselves up. They smile, their lips purple-stained. I remind them that they should be digging for tubers.

They have already forgotten the hunger of the long summer when the hot sun withered everything and they were too weak to play. Two women died over the last season. One was my mother. We have to get the very young ones to help. The holes they dig are never deep enough, but they make a start, and then I can dig down to find tubers.

And when the rain finally came, there was too much of it. The wheat and other grasses are ripening at last, although they are sparse and we must be careful not to harvest the heads of grain when they are too young.

We have walked far, and now we have come to a place where there are plentiful grasses, and trees with fruit we can eat.

Today we arrived at a camp that we visited last season. It is a good place with a stream. There is a dead tree that burns well. The hearth we built last time needed some repair, but I have a cooking fire blazing. The men are returning from their hunt. They have been successful. We will eat meat tonight.

As I fetch water from the stream, I notice a little patch of wheat grass growing. I stand staring at it, wondering why it is denser than the other grasses.

"That is where you tripped and fell last year." It is one of the boys. He is laughing as he recalls me lying face down in the mud, the basket of grain I was carrying spilled all over the ground. "Your mother scolded you." His smile fades.

He is right. My mother was angry with me for spilling the grain that was a day's work to pick. I collected up as much of the spilt grain as I could, grain by grain. She made me keep at it until it was dark.

I kneel down and examine these grasses. I can see that each plant has emerged from a seed. One seed, one plant. The grain I left behind last year has sprouted here, near the stream. I have never thought about how the grain grasses continue to grow, year after year.

I tell the women, but some of them know this already. They have seen the grasses emerging from the grains. I have realised something that they have not considered.

"We could spill more grain here," I say to the women.

They laugh at me, but I was not intending to joke.

"When we have plenty, we can scatter the seeds in good places like this, near water and wood. And when we come back the next year, the grasses will be ready for us to harvest."

"If another clan has not found them first," one of the men says. "Then you will have benefited them and not us."

That is true. I would not want to waste my time feeding another clan.

I have another idea. "This is a good place," I say. "Why don't we stay here? Not for a few weeks, but for a whole year. We could scatter seed and stay to watch over it, bring water from the stream if the rains are late. When the seed sprouts and grows, we can guard it, stop others from picking the ripe grains."

The hunters look at me, puzzled. "We cannot remain in one place. It is … not natural. We must move, follow the animals."

We have stayed at the Place of the Spilled Grain for several weeks. It is time to move on. I am sorry to leave the good hearth, the dead tree and the stream. As the women pack up our baskets and grinding rocks and sickles, I wade across the stream to a spot that gets good sun. In the middle of some grasses that bear seeds we cannot eat, I clear a patch. I have a handful of the grain that I collected today and I sprinkle it on the bare earth. I spread out the grains, so that the plants will have room to grow.

Perhaps next year, if we come back, if no one else has seen the plants, there will be grain growing, ready for us to eat. Perhaps then the other women will see that we could grow our own grain and protect it from others. And we could stay here at the Place of the Spilled Grain.

Unnamed girl, aged 12

It wasn't until about 4 million years ago that human-like creatures, walking on two legs, evolved. It took 3.5 million years for our genus, *Homo*, to evolve from these prehumans, and it was only about 100,000 years ago that modern human beings, *Homo sapiens*, appeared in Africa.

As humans evolved from prehumans to *Homo sapiens* over millions of years, the way these prehistoric humans lived did not change very much. They were hunter-gatherers, living a largely nomadic life as they moved from place to place in search of plants and animals to eat. They cooked their food, but the small amount of greenhouse gas emissions produced by their wood or peat fires was tiny among the emissions from natural events, such as occasional volcanic eruptions and forest fires started by lightning. These early people had no influence on Earth's climate.

THE FIRST FARMERS

Around 12,000 years ago, some humans came up with a radical idea. Instead of being constantly on the move, searching for food plants and hunting wild animals, they thought it would be easier if they stayed in one place, where they could grow their own grain, and tame and

Agriculture is thought to have begun in Mesopotamia, an ancient land that occupied the area from the eastern Mediterranean to the Yellow River Valley in northern China. It began independently, thousands of years later in South America, Africa and New Guinea.

breed wild goats, sheep and cattle to provide them with meat. These people were the first farmers.

They cut down trees to clear fields for growing crops and building pens to enclose their animals. Over a period of thousands of years, the concept of agriculture gradually spread around the world. By 5000 years ago, it had become common practice throughout Europe and South-East Asia.

HUMAN INFLUENCE

Some people believe that human activities began affecting climate a long time ago. American paleoclimatologist William F Ruddiman has concluded that humans actually started to alter the natural level of greenhouse gases back when we stopped being hunter-gatherers and began practising agriculture.

Paleoclimatology is the study of changes in Earth's climate over the entire history of our planet.

Dr Ruddiman discovered evidence that between 5000 and 8000 years ago there was a change in the expected pattern of carbon dioxide and methane levels in the atmosphere, which normally kept pace with the slow swing from warm to cold periods over hundreds of thousands of years. Instead of continuing to fall, very gradually moving the world into another ice age, the levels of these greenhouse gases started to rise again.

Ruddiman believes that these increases in carbon dioxide and methane can be explained by the development of agriculture. The main influences were clearing forests and irrigating fields to grow rice more efficiently.

Some scientists have conducted research that supports this theory. It is possible that humans have been affecting the climate for thousands of years.

PALEOCLIMATOLOGIST

William F Ruddiman (1943–) was born in Washington, DC. When he was at university, during a paleontology class, the lecturer showed his students a vial of tiny microfossils of sea creatures, each the size of a sand grain. The lecturer explained how those minute fossils could be used to reconstruct the cycles of ice ages. Ruddiman was so intrigued, he decided that would be his field of study.

As he was about to retire from his position as Professor of Environmental Sciences at the University of Virginia, he came up with the idea that humans had affected climate for thousands of years. He published his hypothesis in a scientific paper in 2001, knowing that it would be controversial. He is still conducting research on this topic.

DEFORESTATION

Trees and plants remove carbon dioxide from the air during photosynthesis. The carbon dioxide is converted to carbohydrates, such as sugar and cellulose, which the trees use as they grow. Throughout its life, each tree continues to take carbon dioxide from the air and store the carbon within it.

People around the world still cut down the Earth's forests to provide fuel, to make space for agriculture, or to harvest the wood for making things, just as those early humans did. When trees are cleared on a massive scale we call it deforestation. When wood is burned, all the stored carbon is converted to carbon dioxide and released back into the atmosphere in the smoke. If the trees are left to rot, the carbon is released more slowly in the process of decomposition, but it still ends up in the atmosphere as carbon dioxide. A reduction in the number of living trees means that less carbon dioxide is taken up in the process of photosynthesis, leaving more carbon dioxide in the atmosphere.

Around 130,000 square kilometres of forest (equal to the size of England) are cut down every year.

The amount of deforestation has increased as the population of the Earth has grown. The Domesday Book contains information from a survey of England and Wales ordered by William the Conqueror in 1086 CE. It records that 85 per cent of the land had already been cleared to grow crops and keep animals to feed the population. Around the same time in China, a scholar named Shen Kuo wrote about his concerns that too many trees were being chopped down.

There is no dispute that deforestation affects climate change. The debate is about when it started to alter the level of carbon dioxide in the air. Eight thousand years ago the population of the world was somewhere between 5 and 14 million, the size of many cities in the world today. Could the amount of forest cleared by so few people have altered the amount of carbon dioxide in the atmosphere?

"Pine forests in Chhi and Lu have already become sparse ... All the woods south of the Yangze and west of the capital are going to disappear in time if this goes on ..."
Shen Kuo, Chinese scientist and poet, 1088 CE

Land clearing encroaching into the Amazon rainforest.

Deforestation in the Amazon rainforest.

GODS AND DRAGONS

All early humans were at the mercy of the climate conditions wherever they lived (as we are today). They believed their gods and mythical creatures controlled the weather, which was sometimes life-giving, and at other times life-taking. They worshipped weather gods – Thor, the god of thunder in Norse mythology; Ra, the Sun god of the ancient Egyptians; Indra, the god of rain in the Hindu religion – and gave them offerings in the hope they would send rain to water crops, but not so much rain that there were floods.

Ancient Chinese people thought dragons were responsible for making rain. They believed the dragons slept in deep water during winter. People held noisy ceremonies to ensure these beasts woke up in time to bring the spring rains.

ANCIENT SCHOLARS

Some ancient Greek and Roman scholars were very interested in the weather. Aristotle (384–322 BCE) believed that changes to climate and landscape took place "over periods of time which are vast compared to the length of our life".

"For I have found that many authorities now worthy of remembrance were convinced that with the long wasting of the ages, weather and climate undergo a change ..."

Columella, Roman writer, c. 50 CE

Columella (c. 4–70 CE), a writer on agricultural matters, noted farmers believed that over a long period of time weather and climate changed. A writer called Theophrastus (c. 371–c. 287 BCE) had observed that clearing woodland made the climate warmer. Plato (c. 424–c. 347 BCE) wrote about the effects of cutting down trees that grew on mountains.

These scholars were writing about the changes to conditions in their local areas, but they were beginning to realise that climate changed with time, and that sometimes it was because of human activities.

CITIZEN SCIENTISTS

For centuries people around the world have recorded observations of their local climate – farmers noted when crops ripened each year, nature lovers wrote down what plants were growing where, diarists recorded rainfall and temperature.

One English family, the Marshams, was fascinated by their local environment. They were keen observers of the natural history around them. They recorded the dates of events that indicated the arrival of spring, such as the first sighting of a particular yellow butterfly, when the first snowdrops flowered and the first time they heard a frog croak. Six generations of this family recorded these events for 211 years from 1736 to 1947.

In other parts of the world, people have recorded natural

events for even longer. In Japan, cherry blossom festivals celebrate the blossoming of cherry trees each spring, and court diaries have recorded the dates of these festivals for over 1200 years. In China, people have recorded the occurrence of locust plagues for around 3500 years.

Researchers scour museums and libraries for these records and use them to make deductions about what the climate was like in the past. The Marsham records, in conjunction with rainfall and temperature records collected by other citizen scientists, indicate that during the period the family were recording their observations, spring started to arrive a little earlier in England, and it had an effect on flora and fauna. These trends can help predict future reactions of the natural world to climate change.

Citizen scientists still play an important part in climate science today. Sometimes scientists need large numbers of people to help with big tasks, such as counting birds or sieving large amounts of sediment when searching for dinosaur teeth. There are other projects that people can do at home, such as transcribing documents. Scientists ask the public for help and there are often thousands of people who volunteer to assist.

ONE FAMILY'S PASSION

Robert Marsham (1708–1797) was an English country gentleman who had an estate near the village of Stratton Strawless in Norfolk. When he was in his twenties, he started observing the first indications of spring each year. He recorded the dates of 27 natural events, such as plants budding, migratory birds returning and butterflies emerging. He sent his findings to the Royal Meterological Society. He did this until his death at the age of 89. Five more generations of the family took over the task. The last Marsham to note the flowering of the turnip and the first cuckoo call was Robert's great-great-great-granddaughter Margaret Anne Marsham (1890–1954), who stopped collecting the information in 1947 when the society decided it no longer wanted to keep the records. Family members continued to keep records unofficially for at least another decade.

BIG THINKERS

Though a few ancient thinkers had pondered the possibility that human activity could alter Earth's climate, it was not seriously considered until the 1800s. People observed the changes in the weather, but they believed the overall climate of the world was stable.

In Christian countries, people were strongly influenced by the Bible. Most people considered that God was in control of the Earth, not humans. In fact, they believed that God had put humans on Earth to dominate it, to wrest control of it from nature. Some Christians, however, believed it was their duty to care for the wonderful world that God had created for them.

Just as geologists were convincing people that the Earth was not thousands but millions of years old, others were beginning to consider that the Earth had once been a much colder place.

"God made this world to be pleasant to dwell in ... The air, and the sun, and the skies, and the trees, and the grass, and the rivers, they are pleasant things, and we go about spoiling and defacing and deforming them."
William Gladstone (1809–1898), British Prime Minister, 1877

STRAY ROCKS

Geologists were intrigued by particular rocks that they called erratics – large, lonely rocks dotted on a flat landscape or balanced precariously on smaller rocks or pinnacles of earth on a mountain range. These boulders were out of place, made of a type of rock that didn't commonly occur in that area. It was clear that they had come from somewhere else, and the general belief was that they had been carried by the biblical flood that Noah survived in his ark.

People who lived in the Swiss mountains had come to a different conclusion. They noticed that erratics often bore the same long, parallel scratch marks as boulders found in glaciers. They concluded that the rocks had been carried there, not by the flowing waters of a flood, but by the slow-moving ice of a glacier. For these erratics to have been carried from areas where their particular type of rock was common, the alpine glaciers on the peaks of the Swiss mountains must have once been much more extensive.

If there had been an ice age in the past, Earth's climate was not as stable as scholars had previously thought. How this change in climate came about was a complete mystery. Ice-age theories would be debated for the next 150 years.

"The perched bowlders which are found in the Alpine valleys ... occupy at times positions so extraordinary that they excite in a high degree the curiosity of those who see them. For instance, when one sees an angular stone perched upon the top of an isolated pyramid, or resting in some way in a very steep locality, the first inquiry of the mind is, When and how have these stones been placed in such positions, where the least shock would seem to turn them over?"
Louis Agassiz, geologist, 1840

“Erratics” perched on pinnacles of earth, left behind by melting glaciers. The soil around them was washed away, but the boulders protected the soil beneath them from erosion.

4

HEAT AND ICE

19 May 1859

Royal Institution of Great Britain, Albemarle Street, London, England

It's not a bad job working at the Royal Institution, but the hours, spent in a basement laboratory, are long. Once Professor Tyndall is engrossed in an experiment he will never stop, not for food, nor for sleep. Usually, when he's finally finished, I wash the glassware we've used that day and I'm out of the doors as quickly as possible, before he decides to try the same experiment one more time. But tonight is different. I'm not in a hurry to go home. I have come up onto the roof. I wanted to see the sky, even though it's dark.

It's a cold, clear night and London's chimneypots are pouring smoke from household fires into the sky. And though it's dark and the air is far from pure, I can still see through it as

far as the moon and the stars. I reach out to grasp a handful of the night air. But it has no solid shape, no measurable weight. I can't hold it. It appears to be nothing. And yet, it's there. There is oxygen in the air. There is carbonic acid too. I know because I am breathing in and out.

Professor Tyndall spent months preparing the equipment for this current series of experiments, constructing a complicated tangle of tubes and wires. He had to get just the right sort of copper wire and a tube of tin exactly four feet and three inches long. The whole contraption is meant to prove whether gases will let heat travel through. I couldn't see the point. It seemed obvious to me that a cloud of gas couldn't possibly prevent heat from passing through it. The professor, however, has the patience to spend hours, days and weeks to prove something that is obvious to everyone. So we set up the equipment. At one end is the source of the gas to be tested, at the other is a heated plate of copper. Attached to it all is a galvanometer, a device to indicate, by way of a needle on a dial, if the heat is blocked by the gas. The professor first tried oxygen, and the needle on the gauge didn't move. Nitrogen didn't stir the needle, neither did hydrogen. As is his way, the professor repeated the experiments again and again to prove there was no error. The result was the same each time – these gases all let heat pass through them.

Today we had been at it for nearly six hours and I thought I was going to collapse from hunger or thirst.

"Tea, Professor?" I suggested, before he set up yet another repeated experiment.

He didn't object, so I quickly filled a flask with water, lit the Bunsen burner and set the flask on a tripod above it. Professor Tyndall was staring at the flame, deep in thought. I took the opportunity to get out the roast beef sandwich my mother had made for me. When the water boiled, I threw three teaspoons of tea into the flask and stirred it with a glass rod. I was just pouring the tea into the two china cups that Professor Tyndall's housekeeper gave him for that purpose, and adding sugar (we have to manage without milk), when the professor jumped to his feet.

"You have given me an idea, Fletcher," he said. "Let's try the coal gas."

It was the Bunsen burner that gave him the idea, not me. Coal gas is piped into the laboratory for the purpose of heating in experiments (and for lighting when we work into the night). I gulped down my tea, burning my mouth. The professor left his untouched.

We filled the tin tube with unburnt coal gas from the Bunsen burner. It is as clear and colourless as the other gases we had tried, so I was expecting the same result. I took my position next to the galvanometer dial. To my astonishment the needle, which hadn't moved before, suddenly flew across the dial with such speed it hit the stop at the other side.

"Look at that!" I shouted.

Professor Tyndall nodded his head as if he'd known all along that would happen, but I could tell by the twinkle in his eyes that he was delighted with the result.

"Fletcher," he said. "Today we have proved that this gas, though it lets light through unimpeded, blocks heat just as surely as if it were made of solid stone."

He also said that gases like this one, high above us in the atmosphere, stop the heat of the Sun escaping and leaving the Earth freezing cold. It is hard to believe, but our experiment proved it.

Needless to say, we had to repeat the experiment a dozen times or more before he was content there was no error.

Clouds have covered the moon. It's time I went home. Tomorrow we will be testing water vapour and carbonic acid. For once I am quite looking forward to coming to work.

Edward Fletcher, laboratory assistant, aged 16

Professor John Tyndall continued his experiments to measure the heat absorption capabilities of gases over a period of three years. He is recognised as the first person to prove that certain atmospheric gases prevent heat radiating from the Earth's surface escaping into space. And like many other scientists who made the earliest breakthroughs in climate science, he was interested in the theory that the Earth's climate hadn't always been the same.

The understanding of Earth's ice ages and the cycles of warming and cooling that created them are closely linked to our understanding of climate change. During the nineteenth century, many scientists separately studied this topic and their theories and experiments added to knowledge of ice ages and climate. This required the scholars of the day to think beyond the known and imagine the unknown.

In Tyndall's time, there was a great interest in science and many people belonged to scientific clubs and societies. Tyndall gave a series of public lectures to packed audiences. His lectures were attended by celebrities, including Queen Victoria's husband, Prince Albert, and the poet Alfred, Lord Tennyson.

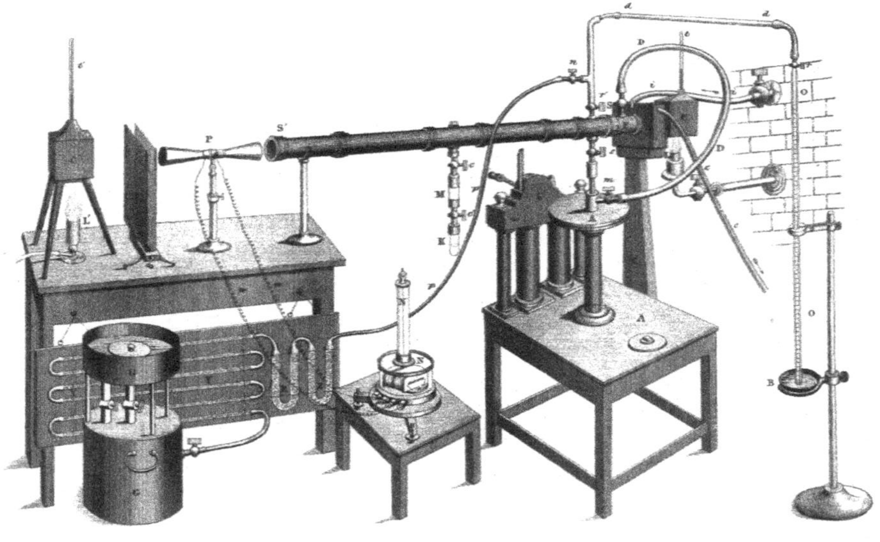

The laboratory equipment John Tyndall used in his experiments for measuring heat absorption by gases.

BUNSEN'S PUPIL

John Tyndall (1820–1893) was the son of an Irish policeman. He worked as a surveyor, but was sacked because he complained about the treatment of workers by the management of the English Ordinance Survey. He went to Germany to study physics and one of his mentors was Robert Bunsen, inventor of the Bunsen burner.

Tyndall was a keen mountain climber, making a number of solo ascents of mountains in the Alps. He climbed the 4634-metre Monte Rosa without a coat, taking with him just half a flask of tea and a bacon sandwich! Therefore, he had firsthand experience of glaciers and erratic boulders, which led to his interest in the theory of ice ages. He was also the scientist who first explained why the sky is blue. Tyndall died after his wife accidentally gave him a lethal dose of chloral hydrate, a drug that he used for insomnia.

ATMOSPHERIC BLANKET

French mathematician Joseph Fourier wasn't interested in ice; he was fascinated by heat and how it is transmitted. And yet his findings were crucial to solving the puzzle of ice ages. In the 1820s, he had pondered why the heat from the Sun didn't just keep heating up the Earth until it was as hot as the Sun itself. He guessed that when the sunlight hits the Earth's surface, it transforms into infra-red heat, which is then radiated away from Earth. Except, there was a second problem. If this was the case, he reasoned, the Earth should be much colder. According to his meticulous calculations (he was a maths genius), the average temperature of the Earth should have been about 30°C colder than it actually is (14.5°C). He checked his figures carefully and knew he hadn't made a mistake.

"The establishment and progress of human society, and the action of natural powers, may, in extensive regions, produce remarkable changes in the state of the surface, the distribution of waters, and the great movements of the air. Such effects, in the course of some centuries, must produce variations in the mean temperature for such places ..."
Joseph Fourier, 1827

At the time, scientists thought that all gases were "transparent" to heat; that is, they allowed heat to pass through them unhindered, just as light did. However Fourier decided there could be only one explanation why the Earth is as warm as it is. The atmosphere acts like a blanket wrapped around the globe, and stops all the heat from escaping. Fourier was the first to express the idea we today call the greenhouse effect. He was also the first to consider that human activities affected climate.

Fourier didn't test this hypothesis himself. And it wasn't until Tyndall turned his mind to the subject 40 years later, after Fourier's death, that the theory was proved. He found that the most abundant atmospheric gases, oxygen and nitrogen, did indeed allow heat to pass through them, but others, carbon dioxide and methane (the main component of coal gas), didn't. Tyndall didn't know it, but he had identified the most important greenhouse gases.

NAPOLEON'S ADVISOR

Jean-Baptiste Joseph Fourier (1768–1830) was a French mathematician and physicist. He was an orphan who almost became a priest, before his interest in mathematics took him on a path of study instead. During the French Revolution he acted as a spy and was imprisoned twice. After his release, he was appointed by Emperor Napoleon as scientific advisor on his campaign in Egypt. There he established educational institutions and dabbled in Egyptology. After three years in Egypt and further work for the French Empire as a prefect and overseer of roads, Fourier returned to his scientific work on heat. Fourier is one of 72 French scientists, mathematicians and engineers whose names are engraved on the Eiffel Tower.

THE BIG FREEZE

Meanwhile, the intriguing idea that in the past there had been an ice age reached the ears of a young Swiss scholar, Louis Agassiz. He travelled to the Jura Mountains to see the erratic boulders for himself and became convinced that they were evidence of an age when ice had covered much of Europe and North America. He presented the hypothesis at a meeting of the Swiss Society of Natural Sciences in 1837. He was the first to publicly raise the ice age theory. There was an uproar as the shocked audience loudly dismissed the theory as the imaginings of an excitable young man.

German botanist and poet Karl Friedrich Schimper (1803–1867) wrote an ode to the Great Ice Age called Die Eiszeit in 1837. It was in German, but it was the first recorded use of the term "ice age".

Geologists were hard at work around the world, studying glacial rocks. In 1871, in Scotland, they found evidence of an earlier ice age. Eventually, geological evidence for five major ice ages was found. Why the Earth's climate had changed from hot to cold and back again remained a mystery.

The most recent ice age was the one scientists focused their attention on, because the evidence in the rocks was more widespread. At first it was known as the Great Ice Age. Later it became known as the Quaternary ice age, after the geological period when it occurred.

The Quaternary ice age started about 2.5 million years ago and we are still in it now.

CELESTIAL MOTION

While geologists were on hands and knees studying the Earth's rocks and glaciers to solve the ice age mystery, James Croll, an amateur scientist from Scotland, was gazing up at the heavens. He stunned scientists in 1875 when he published a book that described the theory that variations in Earth's orbit had altered the planet's climate and caused the Great Ice Age.

Earth's orbit changes shape. Sometimes it is almost circular, at other times it is elliptical or egg-shaped. The angle of tilt of the Earth's axis also alters slightly. The reason for ice ages was simple, Croll said – sometimes the Earth was just further away from the Sun. Therefore, it was colder.

But in the nineteenth century, there was no way to test Croll's ideas. Other scientists politely considered his far-fetched theories, and then dismissed them.

SELF-TAUGHT SCIENTIST

James Croll (1821–1890) was born near Cargill in Scotland. His father's work as a stonemason often took him away from home, so young James had to leave school at 13 to help on the few acres his family farmed. His family could not afford to send him to university, so he educated himself through reading. He developed an interest in science from reading scientific magazines.

Croll tried a number of different professions

including millwright, tea merchant, hotel manager and insurance salesman. An injury to his elbow, caused by a childhood boil, limited the type of work he could do. When he was 38, he was employed as caretaker at Andersonian College in Glasgow. This was his dream job. He had access to the college's library and plenty of time to read its contents. This was where he studied, and wrote his scientific articles and books, including his major work, *Climate and Time, in their Geological Relations.*

HUMAN EFFECT

In the early 1900s, Swedish scientist Svante Arrhenius investigated another of Fourier's theories – that industry would increase the amount of carbon dioxide in the atmosphere.

Arrhenius calculated the effect of increasing and decreasing amounts of carbon dioxide in the atmosphere on the temperature at the Earth's surface. He showed that reducing the amount of carbon dioxide in the atmosphere could have resulted in an ice age. While he was at it, he calculated how much the level of carbon dioxide in the Earth's atmosphere would increase over the next 100 years and how the temperature would increase. He predicted that it would take us 3000 years to double the amount of carbon dioxide in the air.

"... the slight percentage of carbonic acid [carbon dioxide] in the atmosphere may, by the advances of industry, be changed to a noticeable degree in the course of a few centuries."
Svante Arrhenius, 1908

Arrhenius realised that if the atmosphere warmed because of the presence of additional carbon dioxide, there would be more evaporation from the sea and therefore more water vapour in the air. More water vapour (another greenhouse gas) would stop heat escaping from the atmosphere and the air would become even warmer. This was what is now known as a climate feedback loop.

MATHS GENIUS

Swedish physicist and chemist Svante Arrhenius (1859–1927) was a brilliant man who taught himself to read at the age of three. There were no computers in his day, so he made his calculations on paper. It took him 14 hours a day over many months to calculate the effects of increasing and decreasing the amount of carbon dioxide in the atmosphere. This helped him cope in an unhappy period of his life when he was divorced and lost custody of his baby son. He was awarded one of the first Nobel prizes, not for his work on atmospheric carbon dioxide, but for his work in the field of chemistry.

Arrhenius lived in a cold country, so he thought that an increase in atmospheric carbon dioxide making the weather less cold was an excellent idea. He believed it would lead to better plant growth and more food. Scientists were yet to discover that the increase in carbon dioxide could have negative consequences.

TURNING POINT

These nineteenth-century scientists, working in their different fields of interest, were learning about climate in Earth's past, in particular the most recent ice age. Unintentionally, they were also building the foundation of scientific knowledge that would lead to understanding future climate.

At the same time as this scientific endeavour had been evolving, a major turning point in human history had taken place – something that would have an even bigger impact on the atmosphere than the adoption of agriculture. Arrhenius referred to it as "the advances of industry". Today we call it the Industrial Revolution.

5

CHIMNEYS AND MACHINES

1850

Huddersfield, Yorkshire, England

Poor Grandad. I'm watching him sweep the weaving shed. He looks so miserable. He shouldn't have to do the sweeping; it's a job for children. He is a master weaver, or at least he used to be. The manufacturers all knew the quality of his work and called him Mr Taylor. Now his skills are no longer required, because these clanking great machines can do the work of 200 skilled men. I could call myself a weaver now, I suppose, but I have no skills. I just have to watch the machines, make a few adjustments. It's the steam engine that's doing all the work.

There are more than 800 power looms in this one shed. The sound of the clattering machinery is deafening. I'm used to it. Grandad hates it.

He leans on his broom and shakes his head. I know what he's thinking. He used to produce about 24 yards of cloth a week on his handloom. Now one of these great machines, minded by a mere girl like me, can produce more than that in a day.

It's called progress, this manufacturing by machines. The newspapers report on factories with pride. Factories have brought prosperity, or so they say. And each machine is tended by just one or two women. They employ us because they can pay us less. So the prosperity that the newspapers talk about goes to the mill owners, not to the workers.

I used to watch Grandad work when I was little. I didn't feel sorry for him them. He had a handloom in the upstairs room of our cottage. The light streamed in through the long windows and I'd play with strands of wool, marvelling at the way he flung the shuttle back and forth. There was no one telling him what to do. He was his own boss then. He bought the wool himself, already dyed, and Grandma spun it on her spinning wheel. He wove material for fancy goods, such as men's waistcoats. Some were plain and sombre, and others were striped and more colourful. I loved watching the colours change, the patterns emerge – plaids and paisleys. Sometimes he let me help him choose the colours. He laughed when I once chose purple, orange and red together, but he made a few yards up like that. One of the manufacturers took it and made it into smart waistcoats for city gentlemen. It was very popular and after that, I was regularly asked to choose the colours.

This is a worsted weaving mill. There are no pretty colours woven here. The cloth is all browns and greys. Tough, hard wearing. A city gentleman might have a dozen different waistcoats and select them according to his mood. The coats that are made from the worsted in this factory will be bought by working men, who will only own the one, and it will last them all of their life.

Mr Blackburn, the factory owner, used to employ Grandad back in the days before the factories. His son, young Mr Blackburn, is in charge of the factory now. He's a bit sweet on me. He lets me take my dinner break at the same time as Grandad.

"You never knew the town before the factories were built, lass." I don't know how many times I've heard Grandad say this, but I don't mind. I listen as if it's the first time I've heard it. "The Colne Valley used to be a picture, green fields running down to the riverbank and the river slipping quietly by. Then the factories started sprouting up along the river like weeds, all those furnace chimneys pouring smoke. Now you can hardly see the river for the factories, and we only get a glimpse of a blue sky on Sundays."

Grandad wouldn't even have this job sweeping if I hadn't asked young Mr Blackburn if he could give him some work. I have some news for Grandad which will cheer him. "The boss says you can have a few days' work hand-combing next month."

"Aye," Grandad says. "The machine has yet to be invented that can do a good job of combing fine worsted wool. He's not a bad lad, young Mr Blackburn."

He's not so good that he doesn't employ mostly women and children because they're cheaper.

Uncle Ephraim, my Dad's brother, couldn't bear working in a factory. He saved his money and emigrated to Australia. He's gone to the goldfields, in Victoria. Dad keeps saying we should go too, but Grandad is too old for the long sea voyage.

Our dinner break is over. It's time to go back to work.

Anna Taylor, steam loom operator, aged 18

Up to this point in history, humans had only a subtle effect on the Earth's atmosphere, but in a few decades our impact increased dramatically. A single invention changed the world – and changed the atmosphere.

MUSCLE POWER

Since those first farmers had begun to stay in one place, human settlements grew from family groups, to villages and then to towns and cities. People used their intelligence to create tools and equipment to make their daily lives easier, but the pace of change was slow. Even though the wheel was invented around 6000 years earlier, it was still the predominant "machine" in the early eighteenth century.

Though farming techniques had improved, people were still growing their own food – grain and vegetables from the fields; meat, eggs and milk from their animals.

The main industry up until that time had always been agriculture. The power used to plant, plough and harvest was the energy of humans and animals. People planted seed by hand and harvested crops with simple tools, such as scythes. Horses and oxen pulled ploughs, carts and carriages. The only other energy that had been harnessed was the power of water and wind. In mills, these natural forces turned wheels that turned millstones that ground grain to make flour.

The power for manufacturing also came from humans. Everything from cloth and shoes, to buildings and carriages, was made using the skill of human hands. People often did this work in their own homes. Machines had been invented to make manufacturing tasks easier, such as the spinning wheel and the weaving loom, but they were still powered by human hands and feet.

STEAM POWER

Everything changed with the invention of the steam engine, which creates steam, puts it under pressure and converts the heat produced into mechanical energy to operate machinery. Early steam engines were inefficient and used mainly to operate pumps. In 1769, Scottish engineer James Watt, improving on the work of earlier inventors, patented a steam engine. Watt's engine was capable of doing much more than operating pumps. It could turn gears and wheels. This enabled many different machines to be run on the power of steam. The steam engine revolutionised manufacturing. Machines were built that performed tasks much faster than humans could.

But where did all that steam come from?

POWER PRODUCER

As a child, Scotsman James Watt (1736–1819) was sickly and suffered from toothache and migraines. He became an instrument maker manufacturing such things as compasses and quadrants. While mending a steam-driven pump, he realised how inefficient it was. This led him to devise a way to improve the steam engine. Watt needed to compare the efficiency of his steam engines to the previous way of operating machinery. In a particular mill, one of his engines had replaced 12 horses. He worked out how many pounds each horse could pull a distance of one foot in a minute. This became known as horsepower, the unit that we still use today to compare car-engine power. The unit of power for other applications, such as light bulbs, is called the watt, named after James himself. One horsepower equals 745.7 watts.

ROCKS THAT BURN

At least 3000 years ago, people in China discovered the remarkable fact that a particular kind of rock burned. This rock was the final stage in the transformation of all those plants that died way back in the Carboniferous period. Over hundreds of millions of years they had been compressed and transformed into coal.

English people from the thirteenth to the nineteenth century called coal "sea-coal" because it arrived by sea on ships from the northern town of Newcastle.

In Britain, wood continued to be the fuel people used to warm themselves and cook their food, until it became

Perhaps the earliest attempt to control air pollution was a proclamation by King Edward I in 1307 prohibiting the use of coal in limekilns because of the smoke they produced.

scarcer as Britain's forests continued to be chopped down to make way for more farmland. Firewood became very expensive, so British people started burning coal to keep themselves warm as early as 800 years ago.

In the nineteenth century, industrialists looking for fuel to power their steam engines realised that the very thing they needed was right beneath their feet, waiting to be dug up.

Coal was thought to be lucky. People in the nineteenth century (and even the twentieth century) carried a piece of coal for luck. Young girls slept with a lump of coal under their pillow if they wanted to dream of their future husbands.

INFERNAL ENGINES

The cotton industry was the first industry to use James Watt's invention. Huge weaving looms powered by steam took the place of handlooms.

Factories were built to manufacture other goods – everything from tools to cooking pans. Other factories made the nuts and bolts, gear wheels and levers that were required to make even more factories.

Transport was the next thing to run on steam. Construction of railways in Britain began in 1825. The locomotive engines that were soon pulling train carriages were driven by steam produced by burning coal. Then ships powered by steam took the place of sailing ships.

> *"[s]moke is the incense burning on the altars of industry. It is beautiful to me."*
> **William P Rend, owner of the Ohio Central Coal Company, 1892**

The Industrial Revolution is the name given to this period of technological advancement that started in England in 1760 and lasted till 1850. Without coal the Industrial Revolution would never have happened.

IRON MEN

The other resource needed to drive the Industrial Revolution was iron – lots of it.

Coalbrookdale in Shropshire, England is considered to be the birthplace of the Industrial Revolution. It is where Abraham Darby I founded his iron smelting company in 1709. Iron had been in production for nearly 3000 years in Britain, since the Iron Age, but it could only be made in small batches in a furnace. There was plenty of iron ore to be mined in Britain, and iron foundries needed fuel to smelt the iron. Charcoal had been used previously, and had already resulted in deforestation. Now iron needed to be produced on a massive scale.

Coal was needed for iron smelting too, although not in its raw form. Coal contains sulphur and other impurities that make the iron brittle. Abraham Darby perfected the production of a new fuel – coke, which is made by heating coal in an environment without oxygen. Coke has fewer impurities than coal, and burns with more heat. Darby was the first to use coke on an industrial scale.

The factory started out making pots and kettles, but more Abraham Darbys, son and grandson of the first, carried the family business into the Industrial Revolution by making the tonnes of iron needed to manufacture locomotives, bridges and railways.

In 1779, the Darbys also built the world's first cast-iron bridge. Coalbrookdale and the Iron Bridge are now part of a UNESCO World Heritage site.

Abraham Darby III was a friend of James Watt. In 1785, he converted his foundry furnaces so that they could run on the new steam engines.

IDYLLIC INDUSTRY

Like the Coalbrookdale iron foundry, the earliest factories weren't in cities, but in country areas near streams and rivers. Initially, they were operated by waterwheels that relied on a source of fast-flowing water. One of the earliest was Richard Arkwright's cotton mill at Cromford in Derbyshire's picturesque Derwent Valley. The main factory was five storeys high and the work was done mainly by women and children.

Despite the rural surrounds, Arkwright introduced the organisation of mass production, such as 12-hour shifts so that the mill could operate day and night. If workers arrived even a few minutes late, they were not allowed to work that day and didn't get paid.

Arkwright was also among the first to introduce steam engines into his factories. Once steam engines powered factories, they no longer had to be near rivers. Towns like Manchester, Leeds and Glasgow became industrial cities where manufacturing was the main commercial activity. Factories became places full of the deafening roar of machinery. The skies filled with smoke from the coal furnaces that produced the steam.

FATHER OF THE FACTORY

Richard Arkwright (1732–1792) was known as "the father of the modern industrial factory system". The son of a poor tailor, his first business was as a barber and wigmaker and he travelled around buying human hair. When he moved into the cotton-spinning business, Arkwright developed a spinning frame driven by water, which produced a stronger thread than other methods. There is some doubt about whether he was actually responsible for other innovations he patented, but he was knighted and then appointed to the office of High Sheriff of Derbyshire. When he died, Arkwright left £500,000 – a fortune in those days.

PROGRESS AND PROSPERITY

As the factories spread across Britain, many people believed that this massive increase in industry was good for the country and its people. There would be more jobs, and the nation would benefit from the sale of manufactured products overseas. It would free poor people from relying on the whim of nature for their food.

> *"What used to be done by hand and used to take months doing is now accomplished in a few instants by the most beautiful machinery."*
> **Queen Victoria, 1851**

To celebrate all this industry and invention, Prince Albert, Queen Victoria's husband, came up with the idea of The Great Exhibition of the Works of Industry of All Nations. In 1851, British people flocked to see this exhibition where their nation's industry and ingenuity was on display. Working examples of manufacturing machinery were some of the most popular exhibits.

The Industrial Revolution spread from Britain to Europe, then to America and eventually the rest of the world. Coal was the source of all modern conveniences in nineteenth-century Britain. It powered manufacturing, it drove trains and ships, it provided heat for the home.

MODERN CONVENIENCES

At the end of the nineteenth century, a second phase of industrial advancement was heralded by the development of a process for the mass production of steel. Steel is an

Though Thomas Edison is credited as being the inventor of the electric light bulb, he improved on the work of many inventors who had been developing them for more than 70 years. He patented his improved electric light globe in 1880. One of the other inventors who came up with a similar idea around the same time was Australian Henry Sutton, from Ballarat, Victoria.

alloy of iron, but it is much stronger, and can be shaped without breaking or cracking. It was in high demand by industry. Coal, in the form of coke, was crucial to the manufacture of steel.

Another wonder of the Industrial Revolution was gas lighting, which in the early 1800s brought light to streets at night for the first time. There was no resistance from the coal industry – gas was manufactured from coal. When electric light globes became the next innovation in lighting in the 1870s, the coal industry was delighted. Electricity was also generated by burning coal.

Ironically, the one industry that didn't change much with the advancements of the Industrial Revolution was coalmining. That still relied on muscle power – men with pickaxes, boys and even girls pushing tubs of coal along tracks deep underground. By the early 1900s, 1 million people were employed in British coalmines, and they were mining more than 200 million tonnes of coal a year.

COST OF PROGRESS

There was a cost to all this progress. Cities became overcrowded with people who had moved from the country to get work in factories. Many of the factory workers worked long hours (up to 16 hours a day, six days a week) in poor conditions. Children as young as five were put to work in factories because they could be paid much less than adults. Spinners and weavers were once sought-after skilled professionals, potters and blacksmiths had been respected artisans. Once these industries became mechanised, unskilled labourers were employed instead. Workers' income fell.

In the 1840s, laws were introduced to reduce working hours to 12 hours a day, and to ban the employment of children under nine. But even these improved conditions are harsh by modern standards. At the end of each day, workers returned to their cramped homes that had no running water or sewerage.

"Seven hundred and fifty people are sufficient to attend all the operations of ... a cotton mill; and by the assistance of the steam engine they will be enabled to spin as much thread as 200,000 persons could do without machinery."
John Farey Jr, engineer, 1827

Industry was having another effect on the residents of London. The air was full of smoke.

Children working in a textile factory during the Industrial Revolution. The young girl crawling underneath the loom while it is operating is a "scavenger", collecting threads of cotton or wool that might jam the machinery.

6

THE BIG SMOKE

8 November 1852
London, England

I thought I knew London. I walk the streets of this city often, and I believed it impossible for me to lose my way. Mother has sent me to Mr Watkins's haberdashery in The Strand, to get ribbon and lace for the girls in the toy-making workshop at the Ladies Guild. There are other haberdashers closer; however, Mr Watkins gives us a generous discount and saves us offcuts and short ends. I've been there a number of times, but today the fog is so thick it has bewildered me.

Though I left before midday, within half an hour it was more like the last minutes of twilight. I walked down Montague Street and although I could see no more than a yard in front of me, I could find my way by the shop windows which are lit by gaslight. I passed Mr Clarke the tailor and my favourite bookshop. It is a longish walk and I would have taken an omnibus, but the horses that pull them are shy of venturing into the mists and must be led by their driver. I have overtaken several of them.

I turn the corner at Great Russell Street and a boy with a blazing torch offers to lead me to my destination, holding out his grimy hand and demanding a half-penny. I ignore him and stride on.

The mist is an ugly brown, nothing like the white faerie mists of the countryside. This stuff has a sulphurous yellow tinge to it, and I cover my mouth and nose with a handkerchief, for fear of breathing in its noxious vapours. The spire of St Clement Danes appears briefly, and I am reassured that I am still heading in the right direction.

Despite this choking fog, Londoners are still burning coal. Inside each building I pass, someone is stoking a fire, for warmth, sustenance or industry, and adding to the fog. The roofs of four- and five-storey buildings disappear into the wreathing greyness above me. I turn a corner and another, and now I do not know where I am. There are no familiar shop windows, no gas lamps, just dark and evil-smelling streets.

The fog dulls all sounds and only a barking cough and the wail of an infant break the eerie silence. I see a smudged orange glow before me, and suddenly I am confronted by a coarse female face and a calloused hand that is thrusting something at me. I feel the heat from her brazier.

"Roasted chestnuts," she says. "Penny a bag."

I shake my head and hurry on.

The fog has turned dark brown. It is a perfect medium for theft and other kinds of lawlessness. Any crime would be concealed. I attempt to retrace my steps, plunging blind into the fog, which is now almost black. Every turn I take only draws me deeper into this veiled world. I am lost in a maze.

A disembodied hand emerging from a coat cuff, frayed and stiff with grease, reaches out and grabs my sleeve. The fog swallows my cry. I am about to struggle to release myself from this grip, but a face emerges from the gloom with a gap-toothed smile.

"Careful, Miss Hill." I realise it is one of the boys from our school.

He points to the ground in front of my feet. It drops away suddenly and, though I cannot see it, I can hear the slap of water. I have somehow crossed The Strand without realising it, and have nearly walked into the Thames.

"Thank you, Frankie," I say. "You have saved me."

I fumble for a sixpence. He holds up his grubby hand.

"No need, Miss." A gust of wind thins the fog. He points a gloved finger. "If you're going home, it's that way."

I can see the glow of a gaslight hovering in the air in the distance. I hurry towards it and grasp its post as if it is an old friend. Through the mist I see a shop I recognise. I forget about the ribbon and lace and turn homewards.

Our capital city is now like Hades. Its inhabitants tolerate these fogs, oblivious to the poisonous vapours they contain. The face of the sun sometimes isn't seen for weeks. Trees and people alike wilt and die from its effects.

Something must be done!

Octavia Hill, social reformer, aged 15

In Victorian England, the foul fumes belching day and night into the atmosphere from factories, furnaces and fireplaces contained tonnes of carbon dioxide. But the people were unaware of the increasing amount of this invisible greenhouse gas that was finding its way into the atmosphere far above their heads. They were only concerned about pollution they could see and smell. The sort that directly affected their health.

BAD AIR

Carbon dioxide isn't the only gas released when coal is burned. Sulphur dioxide and nitrogen oxides such as nitrogen dioxide are also produced. Sulphur dioxide has a sharp, acidic smell that burns the nose and throat. Nitrogen dioxide is a toxic gas with a reddish-brown colour.

"... to pile up one hundred thousand factory chimneys, vomiting soot, to fill the air with poisonous vapours till every leaf within ten miles is withered, to choke up rivers with putrid refuse ... and overhead by day and by night a murky pall of smoke – all this is not an heroic achievement ..."
Frederic Harrison, historian, 1882

Nineteenth-century coal furnaces were not efficient. Coal smoke also contains soot – black particles of carbon that hadn't been completely burned. This other by-product of coal burning hung in the air visible to everyone.

The inconvenience of smoke in the air was a small price to pay for the convenience of warm homes, train travel and plenty of jobs. At least that's what the owners of factories and coalmines said.

A nineteenth-century textile factory.

AHEAD OF HIS TIME

Long before the Industrial Revolution, people were complaining about the unpleasantness of the smoky air in London. As early as 1661, according to a man named John Evelyn, the air in England's capital city was thick with smoke. People were already warming their homes with coal fires, and small industries such as limekilns and blacksmith forges burned coal.

"[London's] inhabitants breathe nothing but an impure and thick mist accompanied with a fuliginous and filthy vapour ..."
John Evelyn, writer, 1661

Evelyn wrote a book on the subject of London's terrible air. He was 200 years ahead of his time claiming that the "coal smoake" was bad for people's health and responsible for "corrupting the lungs". No one took any notice of him.

When the Great Plague reached London a few years after Evelyn's book was published, people were encouraged to build coal bonfires in the streets. It was believed that the smoke would disinfect the air.

VISIONARY GARDENER

John Evelyn (1620–1706) was born into a wealthy English family. He is most well-known for his diary, which he began work on at the age of 11 and wrote until his death. He was a Royalist in the English Civil War, but he travelled to Italy and France to avoid the "hazards" of war. His favourite occupation was gardening and he spent a lot of time on his brother's estate. His book about the smoky atmosphere of London was published in 1661. Evelyn thought that sweet-smelling trees planted around the city might help improve the air.

WORLD'S FIRST POLLUTED CITY

London was the first city in the world to suffer from air pollution. During the 1830s, industry and households in the city burned 2.5 million tonnes of coal a year. Fog in London was becoming thicker and more prevalent and it was a different colour to previous fogs, but at first people didn't make the connection with the smoke. London became known for yellowish fog "the consistency of pea soup." Eventually people realised that a natural fog was made up of water vapour and these fogs that contained smoke needed another name.

A campaigner against coal smoke, H A Des Voeux, is believed to be the person who combined the words "smoke" and "fog" to come up with "smog".

Most Victorians thought that smoke was just a nuisance, something that left black specks on the washing, and meant that you had to clean the windows more often and light lamps during the daytime to see what you were doing. However coal smoke wasn't just inconvenient. When rain fell, it cleared the air, but the raindrops absorbed the sulphur dioxide and nitrogen oxides in the air, forming sulphuric and nitric acids. Stonework crumbled under the corrosive effects of this acid rain. Reduced sunlight and acid rain also affected plant growth, which made food scarcer and more expensive.

The current British Houses of Parliament (including Big Ben) were built between 1840 and 1870. London at the time was so polluted that the acid in the air started to eat away at the stonework before the building was even finished.

A LONDON FOG.—DRAWN BY DUNCAN.

In the 1800s, "link boys" carrying burning torches (which would have added to the pollution) did good business charging a farthing (a quarter of one penny) to light people's way home through the smog.

BROWN OR BLACK?

Calls to improve London's air were revived, and in 1853 a *Smoke Nuisance Abatement Act* was passed in parliament, requiring manufacturers to limit the smoke that factories produced. Laws stated black smoke could only be emitted from factory chimneys for ten minutes per hour, but there was no technology to measure air pollution. The officers whose job it was to gauge the level of pollution compared the smoke to a chart with shades of grey, from faint to black, representing smoke density. Then they timed how long black smoke was emitted. Anyone emitting more than the permitted amount was fined.

Some factory owners argued that their smoke was dark brown, not black. Some factories were let off when they claimed that they would go out of business if they reduced the amount of coal they burned. Iron and steel manufacturers, some of the worst offenders, were exempt from fines because the steel industry was considered to be too important to the nation's economy. There were no household restrictions, so Londoners continued to burn coal at home, and there was no reduction in the amount of smoke in London's atmosphere.

"HEALTHY"

Factory owners continued to insist coal smoke was harmless; in fact, some claimed breathing in smoke was good for people because it acted like an antiseptic and the sulphurous gases in coal smoke were like a tonic. Others said that sewers could be cleaned by pumping chimney smoke into them.

Doctors recommended that people suffering from tuberculosis should inhale coal smoke. Some people were concerned that reducing the smoke in the air might actually be bad for people's health, as the smoke fumigated the air, killing whatever it was that caused diseases such as malaria (they didn't know about bacteria and viruses yet). But eventually doctors realised that smoke was causing illness, particularly lung disease, just as John Evelyn had claimed 200 years earlier.

DEADLY

A pamphlet called *London Fogs* was published in 1880 by amateur meteorologist Rollo Russell. It was a bestseller. Russell wrote that London's smog was killing people and that in one three-week period 2000 deaths were attributable to it. One of the main causes of death was respiratory disease. Russell also said that the constant fog and resultant lack of sunlight was causing depression that led to people drinking too much alcohol. He suggested

> *"We must reckon a large annual loss of life from the perpetual presence in the London atmosphere of smoke and soot, blocking up the air passages and irritating the mucous membrane so as to lead to consumption, lowering the vital energy, depressing the system both by the impurity of the air breathed and by the deprivation of light ..."*
> **Rollo Russell, meteorologist and writer, 1880**

a number of ways to reduce the amount of smoke produced by households. He said people should burn smokeless fuels, such as coke and gas, and use only fireplaces and stoves that were efficient. Inefficient stoves should be taxed.

Scientists developed ways to measure how much tar, ash and soot were falling to the ground from the smoke every year. In London, the estimate was a staggering 353 tonnes per square mile. But harmful gases, such as sulphur dioxide and nitrogen oxides, still weren't monitored.

FIRST ENVIRONMENTAL GROUP

Since neither factory owners nor the government seemed concerned about air quality, some Londoners started to take action themselves. In 1880, a group of concerned citizens – led by social reformer Octavia Hill – formed the Fog and Smoke Committee to tackle the problem. It was the first environmental group to focus on pollution.

> *"... these gardens for the people everywhere look reproach on me when I think of England, and every tree and creeper and space of green grass in [Nuremberg] reminds me of our unconsumed smoke and how it poisons our plants, and dims the colour of all things for us."*
> **Octavia Hill, social reformer and smoke abatement campaigner, 1880**

In 1881, the Fog and Smoke Committee organised a Smoke Abatement Exhibition with an emphasis on improvements in household heating methods that would reduce smoke. Thousands of people flocked to see fire grates that used smokeless fuels or reintroduced the smoke to be burned more thoroughly by the hot coals. They also demonstrated newfangled gas stoves.

CLEAN AIR CRUSADERS

Octavia Hill (1838–1912) was an English social reformer who worked all her life to improve housing and education for London's poor, often driving herself to a state of collapse. She started helping the poor when she was 14, supervising girls at a ragged school. During a visit to Nuremberg in Germany she was struck by the better living conditions for poorer people; in particular, she noticed the trees and green spaces in the city. When she returned to London she joined forces with Ernest Hart (1835–1898), a Jewish doctor specialising in eye disease. Together they formed the Fog and Smoke Committee in an attempt to improve London's air. Octavia also founded the National Trust to preserve green open spaces for public use.

OUT OF SIGHT

Londoners were keen to get rid of the smog that fouled their city – but reluctant to give up their familiar coal fires. They didn't like using the smokeless fuels that were harder to light and burned with a different coloured flame that didn't look as cosy as the coal fires they were used to. A respected scientific journal claimed a good fire had to be "open, pokeable and companionable" and that coal fires in the home were responsible for the British having healthy complexions and good eyesight!

"... and we believe it is to the use of open fireplaces that the general freshness of complexion of the inhabitants of these islands, and the absence of the use of spectacles among the young, are in a very large measure to be attributed."
Report on the Smoke Abatement Exhibition, 1882

Gas companies heralded gas as the fuel of the future. Households using gas and coke produced much less pollution, and the air in the City of London began to clear. But the problem of pollution had not been solved.

The gasworks that burned coal to produce gas and coke were still polluting, pouring out toxic smoke that contained methane and carbon monoxide, while waste products flowed straight into rivers and streams, or soaked into the earth. In the mid-1800s, every town and city in Britain had its own gasworks. London had 13, but they were in poor areas where people, grateful for the work they provided, didn't complain about the

When London's Millennium Dome, which housed an exhibition to celebrate the new millennium, was built on the site of an abandoned gasworks, 200,000 tonnes of contaminated soil was removed at a cost of £185 million.

shocking conditions for workers or the pollution in their neighbourhood.

KILLER SMOG

World War II brought coal rationing and wartime dangers, more immediately threatening than pollution. Then in December 1952, as coal rationing began to be lifted, smog returned to London. This one was so dense all transport stopped. Football games were cancelled. More seriously, a subsequent inquiry revealed that more than 4000 people had died from the effects of this smog. As levels of pollutants were still not monitored, no one knew exactly which part of the smog was responsible for the deaths. The inquiry concluded that it was the coal fires in people's homes that were the problem. They burned coal inefficiently, creating twice as much smoke per tonne as industry did.

During the 1952 smog, a performance of *La traviata* at Sadler's Wells Theatre was cancelled after the first act because the theatre was full of smog.

Four years later, the *Clean Air Act* was passed in parliament. It was the first effective law to control air pollution. Octavia Hill and the Fog and Smoke Committee would have been proud. The Act made it compulsory to use the type of smoke-reducing technology that they had displayed at their Exhibition more than 70 years earlier.

CLEARING THE AIR

Smoke Control Zones were designated throughout London where only smokeless fuels, such as coke and black, hard coal called anthracite, could be burned. The level of smoke in London's air continued to gradually fall from almost 300 micrograms per cubic metre in the 1950s to below 100 micrograms in the mid-1960s.

In 1965, natural gas was discovered in the North Sea, off the coast of Scotland and England. Gas became cheaper and was no longer produced by burning coal. People reluctantly gave up their coal fires for gas heaters. By 1969, half of Greater London was designated a "smokeless" zone.

It had taken more than 100 years, but smoke was no longer a problem in London. By 2000, the level of smoke in London was virtually zero.

7

BLACK GOLD AND TIN LIZZIES

12 May 1908

Masjed Soleyman, Persia

I do not understand the pale foreigners. They do not like being in my country, so why don't they go home? They hate the hot sun that burns their white skin, and they don't like the cold of the desert nights either. All they do is complain.

The leader of the foreigners is a funny-looking man. His face is red, his beard is orange and his eyes are blue. I call him Mr Rainbow. He wears short trousers that only reach to his knees and his pale hairy legs poke out of them like sticks. Instead of a turban he wears a stiff hat like an upturned bowl.

He says they are looking for "oil". That is what they call the thick black stuff that oozes out of the ground in some places. It is useful for sealing water skins and it can be made into the liquid that we use in our lamps. But there is enough for those purposes if you know where to look for it. The foreigners say they want a lake of oil. I don't know why they need so much.

My family herds goats and we are camped near where the foreigners have built their oil rig, an ugly thing made from wood and iron. For many months, a noisy engine has been running day and night, making a drill burrow deep into the ground. The foreigners have lots of money and some Bakhtiari men work for them.

I have learned some of the foreigners' words and I know that all their workers wish they could go back to their homeland. They complain about the food. To earn a little money, I have tried to sell them goats' milk and dates, but they do not want to buy these things. When they tasted the milk, they spat it out.

We Bakhtiari eat goat, rice and lentils. Although it is good food, the foreigners will not eat it. Their food comes out of small metal cylinders sealed at both ends, which they call tins. They open one end to get the food out. These tins travel on ships across the sea from the foreigners' homeland, far away. Some of the tins contain beans in a strange red sauce; others pale meat that they eat cold. One of the foreigners gave me some to try. I would not eat it because it smelt bad. The only food of ours that the foreigners will eat is bread, though they complain about that too, saying it is not like the bread of their homeland. Some other metal tins contain a thick greenish stuff that looks like slime from a pool that is nearly dried up. They spread it on bread. "Gooseberry jam" they call it. I have tasted it. It is very sweet. I do not like it.

One of the foreigners is called Wilson. He is the only one who likes our food. Mr Wilson says he is too busy to prepare meals every day, so he wants me to cook for him. I will stay while my family's goat herd is grazing nearby.

Mr Rainbow has been looking miserable ever since a messenger brought a sheet of yellow paper from Basrah. He says he must go home. The other foreigners are very happy to be leaving, but Mr Rainbow is sad. I think he is a little crazy. I have seen him muttering to the rig. Telling it to work harder.

26 May 1908, 4.30 am

Loud shouting wakes me from a dream of climbing a palm tree to pick dates. I think another tribe must be attacking us. Then I see it is the men who work the oil rig through the night who are making the noise, and I realise these are not shouts of fear, but of happiness. In the grey light before dawn, I can see that they are all dancing, foreigners and Persians together. Hats and turbans are thrown in the air.

"Oil!" The foreigners shout. "We've struck oil!"

The oil is shooting into the sky, three times the height of a palm tree. It is like a black fountain. And the oil is raining down on everyone. Raining on me too. We all look the same now. Our faces are black, shiny and smiling.

Rostam Moshiri, goat herder, aged nine

The battle against air pollution was being won in towns and cities, but the global fight against atmospheric pollution was just beginning. Coal consumption, though out of sight of most people, continued to grow. And there were other fossil fuels besides coal hidden underground that would help feed the growing hunger for energy that was spreading through the developed world in the early 1900s.

BLACK GOLD

"This liquid is fatty and sticky like the juice of meat. It is viscous like congealed grease. If one sets light to it, it burns with an extremely bright flame. It cannot be eaten."
Description of oil written during the Eastern Han dynasty (25–220 CE) in China

Those masses of marine plankton that died in the oceans millions of years ago had slowly converted to oil. Throughout history people have found uses for oil wherever it has seeped to the surface, unaware of the vast oil deposits beneath them. Ancient Egyptians used oil to seal wounds. The ancient Chinese used it to grease cart axles and make better ink. In Persia, a scholar named Muhammad ibn Zakariyā Rāzī (865–925 CE) distilled oil to produce kerosene to burn in lamps. Persians also wrapped their arrow tips in oil-soaked material and set them alight.

The first commercial oil well was in Pennsylvania, USA. It was operated by E L Drake who struck oil in 1859. Like the ancient Persians, he used it to make kerosene, which was still a popular fuel for lamps. Oilfields were found in other parts of America, with some of the largest finds in Texas and California.

BOOM AND BUST

Drilling for oil was a lucrative business, which made the fortunes of many Americans, including millionaire John D Rockefeller. But in 1879 another American, Thomas Edison, patented the electric light globe. Electric lights didn't smell bad like the lamps used at that time, which burned whale oil and kerosene. It wasn't long before electricity cables were connected to communities so that people could light their businesses and homes with these modern inventions.

When oil production began in the 1860s, oil and petroleum products were stored and transported in wooden barrels of different shapes and sizes. By the early 1870s, the 42-gallon barrel had been adopted as the standard for the oil trade. The standard barrel size used by other industries was 40 gallons. The extra two gallons were added to make up for evaporation and leakage of the oil during transportation.

The market for kerosene disappeared. The price of oil fell from a high of $10 a barrel in 1860, to less than a dollar a barrel. It looked as if Rockefeller and everyone else who had invested heavily in oil would lose their oil wealth. Fortunately for them, another invention came along that relied completely on an oil product – the automobile.

STANDARD OIL COMPANY

John D Rockefeller became interested in oil at the age of 26. With five business partners, he founded the Standard Oil Company in 1870. Rockefeller then went about taking over the US oil market. Standard Oil bought up struggling oil companies, and in less than ten years owned more than 90 per cent of America's oil production. The company became a ruthless and secretive monopoly controlling the railway lines and the pipelines that transported oil and making huge amounts of profit. Standard Oil spread overseas and became one of the first multinational corporations.

The US Department of Justice accused Standard Oil of unfair business practices and demanded that the monopoly be broken up. In 1911, the US Supreme Court ruled that Standard Oil was to be divided into 38 parts. Three of those parts were Standard Oil New Jersey, Standard Oil New York and Standard Oil California. These later became Exxon, Mobil and Chevron respectively, companies that still dominate the oil industry today.

CARS TO THE RESCUE

Though historians say the Industrial Revolution ended in 1850, invention and industrial development didn't stop then. Steam engines could be adapted for use in a large vehicle like a train. However when the need for a personal road vehicle arose, the steam engine was unsuitable.

It would take around 25 tonnes of prehistoric plant life to create every litre of petrol we use today.

Inventors had been developing the internal combustion engine throughout the nineteenth century, and had tried gas and electricity to run cars. Petrol had previously been a waste product from kerosene manufacture, and it won out as the best fuel for road vehicles.

Though a number of men were building horseless carriages in the late nineteenth century, German Karl Benz (1844–1929) is recognised as the inventor of the first car. His Motorwagen looked more like a tricycle. It had three wheels, a two-horsepower motor and one gear. There was no steering wheel. The vehicle was steered with a handle.

Benz & Cie. briefly led the world in car manufacture, but cars quickly evolved from the fragile Motorwagen to the more robust Ford Model T. The US Ford Motor Company began the mass production of cars in 1913 and soon took over the title of world's leading car manufacturer.

Before long there were about a million cars on US roads, and they all ran on petrol that could be made cheaply from American oil. Automobiles saved the American oil industry. Soon other forms of transport were also using the products of oil refining – diesel trains, aircraft and eventually diesel-electric ships.

BERTHA BENZ (1849–1944)

German Bertha Ringer was a woman with the mind of an engineer, who assisted her fiancé Karl Benz as he developed the first car. Bertha invested her family's money in the venture. After their marriage in 1872, laws prevented her from further financial investment, but they couldn't stop her marketing his first car. In 1888, without Karl's knowledge and with her two teenage sons on board, she drove the Motorwagen 194 kilometres to her mother's house. Before that distances travelled by early cars had been measured in metres. Bertha had made the first ever long-distance car trip. Her history-making journey was a great publicity stunt that proved the capabilities of the Motorwagen. The problems encountered en route (all solved by Bertha and the boys) enabled Karl to make improvements.

NEW OIL, OLD COAL

The oil discovered in America was being used up at a breathtaking rate. But discovery of oil fields in other parts of the world meant that no one had to worry about running out of oil. Not yet.

In the first half of the twentieth century, vast resources of oil were discovered beneath the desert sands of the Middle East. The oil well at Masjed Soleyman, in what is now Iran, was the first oil find in the Middle East. It belonged to the British, who had purchased a concession allowing them to drill for oil in most of southern Persia for 60 years. Oil was discovered in Iraq in 1927 and then in the other states around the Persian Gulf. These oilfields were all operated by British and US oil companies.

The first oil find in the Middle East was financed by William Knox D'Arcy (1849–1917). He was an Englishman who emigrated to Queensland, where he acquired shares in a gold mine and became a millionaire. He invested his money in oil exploration in Persia. After years without success, he was almost bankrupt when his team struck oil at Masjed Soleyman. He became a millionaire all over again.

The Middle Eastern states eventually took control of their own oil, forming the Organization of the Petroleum Exporting Countries (OPEC) in 1960. Other oil-producing nations joined and OPEC now has 12 member countries – Algeria, Angola, Ecuador, Iran, Iraq, Kuwait, Libya, Nigeria, Qatar, Saudi Arabia, the United Arab Emirates and Venezuela.

The Middle East has only two per cent of the world's oil wells, but the oilfields are so big that they provide for around a third of the world's oil needs. It is estimated that OPEC countries currently hold 81 per cent of the world's oil reserves – 1206 billion barrels.

The discovery of oil and the end of steam-driven transport could have meant the death of the coal industry. It didn't. Electricity was still predominantly produced by coal-fired power stations.

And we needed a lot of it. As the twentieth century progressed, every innovation seemed to run on electricity – the phonograph, radio, television. Advertising convinced people that no household could manage without the latest electric appliances, such as hairdryers, refrigerators and washing machines.

KNOCK, KNOCK

Cars were continually being improved to attract more buyers. One problem was engine "knock", caused by incorrect combustion of petrol and resulting in inefficiency and engine damage.

In 1916, Thomas Midgley Jr, an American mechanical engineer working for the car manufacturer General Motors, was given the task of finding the right chemical to add to petrol to stop the problem. In 1921, he decided a chemical called tetraethyl lead (TEL) was the right chemical for the job. It was very successful, leading to improved fuel economy and engine performance. General Motors joined forces with Standard Oil, and petrol containing TEL went on sale in 1924. The use of leaded petrol soon spread around the world.

But there was a problem. TEL was toxic. The lead it contained damaged humans' organs and nervous systems.

DEADLY SOLUTION

Midgley and General Motors were aware of the serious health effects of lead. Dozens of men working in the manufacture of leaded petrol had died of lead poisoning; others became violently insane.

Hallucinations of insects flying around workers' heads were so common that one oil refinery manufacturing leaded petrol was nicknamed "The House of Butterflies".

At a press conference, Midgley poured what he claimed was TEL over his hands and inhaled the fumes, to prove it was harmless. This was despite the fact that he had himself suffered from lead poisoning during his research and had been forced to take an extended break from work to recover.

It wasn't only workers who were affected. Everybody was at risk. The manufacturers of leaded petrol knew that car exhaust fumes contained lead. People on the streets breathed the lead in and it entered their bloodstreams, leading to serious and widespread health issues. General Motors, and the oil companies that added lead to their petrol, including Standard Oil, had conducted research on lead and health. They were aware how dangerous it was, but they didn't make their findings public.

"Present day civilization rests on oil and motors ... We do not feel justified in giving up what has come to the industry like a gift from heaven on the possibility that a hazard may be involved in it."
Frank Howard, Standard Oil executive, 1925

HELLO, CARBON MONOXIDE

Cars contributed to another environmental problem that everyone could see and smell – air pollution. Twentieth-century air pollution was different to the stuff that troubled people in Victorian England. Industry had been forced to clean up its operations, and though factories still emitted pollutants, the major source of air pollution in cities had become car exhaust fumes.

As well as lead, car exhaust contained dangerous waste products – nitrogen oxides, hydrocarbons, carbon monoxide and particles from unburnt fuel. They all had harmful effects on human health.

RETURN OF THE SMOG

In the twentieth century, Los Angeles in the USA was the city most affected by smog.

Just as in the previous century, smog was at first considered to be nothing more than a nuisance. Then in 1943, Los Angeles had its first severe smog day that made residents' eyes and throats burn. This was in the middle of World War II and some people thought the smog was the result of a gas attack by the Japanese. It was bad enough to convince authorities that something had to be done.

Other results of the smog were observed – crops looked bleached and withered, tyre manufacturers noticed that rubber deteriorated faster in Los Angeles than in other cities, doctors began to realise that smog was affecting the health of the city's inhabitants. The first authority in the USA to control air pollution was formed in Los Angeles in 1947. They blamed the smog on emissions from oil refineries and people burning rubbish in their backyards. One scientist disagreed.

THE CULPRIT

In 1949, farmers around LA were complaining that the smog was making their crops wilt. A chemist called Arie Haagen-Smit was called on to find out why. He thought he detected a faint bleach-like smell in the smog. This led him to believe that the pollutant in LA smog causing all the damage was ozone.

This was a surprise, as ozone is not directly produced by industry or by cars. It is created in a chemical reaction caused by sunlight acting on hydrocarbons and nitrogen oxide. Both of these chemicals were waste products pouring from the exhaust pipes of all the cars that clogged LA freeways. Though ozone wasn't directly produced by the combustion of petrol, it was still car exhaust fumes that caused smog.

Los Angeles, famous for its sunshine and freeways, was the perfect place for smog to be created. And it was the ozone in the smog that was causing eye and throat irritation, and harming plants. This type of smog is called photochemical smog.

A SENSITIVE NOSE

Arie Haagen-Smit (1900–1977) was born in Holland where his father was the chief scientist of the Royal Mint. As a child he was interested in maths and taught himself calculus, but at university he studied organic chemistry.

He moved to America in 1937 and isolated the chemicals that made radish leaves grow and those that made the eyes of the fruit fly red. His main scientific interest was in the chemistry of essential oils in plants. He was trying to identify the chemical responsible for the smell and taste of pineapples when he became involved with smog research.

His work on smog was not only in the laboratory. He became chairman of the California Air Resources Board, and used his skills as a diplomat and negotiator to fight for legislation to control smog.

CLEANING UP

Car manufacturers tried to argue that the ozone came from the stratosphere, and was nothing to do with cars. Oil companies banded together and tried to discredit Haagen-Smit and his findings. They were not successful. In the 1960s, California introduced its first laws to reduce car exhaust emissions.

The removal of lead from petrol finally began in the

Leaded petrol was phased out in the USA by 1996. In Australia, it wasn't completely withdrawn from sale until 2002. There are a few countries that still use leaded petrol today.

1970s. This wasn't because the car industry had decided it was wrong to inflict poisonous lead fumes on people, it was because the lead was damaging the new catalytic convertors they had been forced to add to their cars by the new *Clean Air Act.*

California has led the world in the production of less polluting cars. In the 1990s, that state set America's strictest standards for cleaner petrol and for low-emission vehicles. In Los Angeles, there are now fewer smoggy days and the severity of the smog has diminished dramatically, but Californians cling to car use, and the number of kilometres travelled continues to soar. There is still smog in LA and in other cities around the world where cars and trucks provide the main form of transport.

Even with lead removed from petrol, and ozone produced by car exhaust limited, our modern cars and trucks are still responsible for the production of the most dangerous greenhouse gas – carbon dioxide.

Transportation produces 13 per cent of the world's greenhouse gases.

8

SOLID SCIENCE

13 March 1993

Hulu Cave, Tangshan, Jiangsu Province, China

All I can see is a small circle of light above me, blindingly white, surrounded by blackness.

A few moments ago, I was playing soccer with my friends. We call ourselves the Tigers. We were playing against the Roosters. They aren't as competitive as we are. We were halfway through a match and I had the ball. I kept it just in front of my right toes, zigzagging around the opposition whenever they tried to take it from me. When I was about two metres from our goal (made from three bamboo canes), I kicked the ball hard and it made that satisfying "doof" sound. The ball flew in a perfect straight line towards the goal. The boy who is the Roosters' goalkeeper looked terrified and lunged sideways. He wasn't trying to stop the ball; he was trying to avoid it hitting him. Another goal to the Tigers! I leaped up and punched the air, as I had seen soccer players on television do when they celebrate a goal. My feet hit the ground, but they didn't stop; they kept going. My teammates and opponents all stared at me with their mouths gaping, as a hole opened up beneath me and I disappeared into it. I didn't scream. I didn't want my friends to think I was afraid, and anyway, the rush of air took my breath away. I fell for so long, I thought I must be falling to the centre of the Earth. Then I hit rock. I had never felt such pain. I clenched my teeth, still determined not to scream.

A head appears in the circle of light above me.

"Tengfei!" a frightened voice calls down. "Are you dead?"

"No," I manage to call back. "Not quite."

It is one of my teammates. "Longwei has gone to get help," he says. "The Roosters have run away home. They're afraid they'll get into trouble."

I turn my head a little and I can make out spikes sticking up from the cave floor all around me. Nothing to be afraid of, I tell myself. They're just stalagmites. I am in an underground cave. Luckily I didn't fall on one of the limestone spikes. It would have killed me.

As my eyes adjust to the dim light in the cave, I can see that there are matching stalactites hanging from above. My left arm won't move. I know it's broken.

I can hear the steady sound of dripping water. The rock is wet underneath me. I am cold. I pull myself up into a sitting position. Every movement means terrible pain in my arm.

"There's something next to you, Tengfei," my friend says. His voice quavers as if he's afraid. He's got nothing to be scared of, safe up there in the sunshine. Then I turn my head the other way. Lying next to me is the grinning face of a skull. I scream.

19 March 1993

My arm is in plaster and I am lying on the sofa watching TV, even though it is a school day. It is nearly a week since my fall, and I am still bruised and sore all over. My mother comes in with a man I don't know.

"This is Wang Yongjin," my mother says. "He's come to see you."

"I am from Nanjing Normal University," the man says.

I think I must be in trouble because I haven't been at school for a week, or perhaps for breaking a hole in the playing field, but Mr Wang smiles at me.

"Your fall has helped us find an important cave," he says. "Those skeletons are extremely old, perhaps 600,000 years. And the stalagmites have been growing for a very long time. They contain information about the past that we hope will tell us about what the world was like thousands of years ago."

I didn't know stalagmites were so important.

"You and your friends are adventurous boys, I expect," Mr Wang says. "I dare say you have explored the countryside around

here. Perhaps you know of other caves like the one you fell into."

I do. I tell him about Linzhu Cave. He is very pleased. He says he's going to come back with a team of scientists from the university to study the caves.

Tengfei Ma, student, aged ten

There have only been professional scientists for about the last 150 years. Meteorologists have recorded weather data for around 130 years. In the early 1900s, climate science was still in the hands of clever amateur scientists. As the twentieth century progressed, it became the province of professional scientists. Some left their desks and laboratories to investigate traces of past climates in the natural world – including stalagmites. Modern communications enabled scientists from around the world to communicate with each other and to work together on projects.

But at the beginning of the twentieth century, there was at least one amateur scientist who was still thinking about the Great Ice Age.

STRETCH, TILT AND WOBBLE

Serbian civil engineer Milutin Milanković believed the amount of sunlight in summer controlled the increase or decrease in the amount of ice on Earth. Over a period of 30 years in the first half of the twentieth century, he expanded his theory and made painstaking calculations to develop a mathematical theory to prove it. He calculated the two factors that Croll had mentioned – Earth's elliptical orbit (known as eccentricity) and the tilt of its axis (obliquity). He added another. There is a "wobble" in Earth's axis that causes it to change the direction in which it is pointing (precession). Each of these variations in the Earth's orbit and rotation results in a slight variation in the amount of sunlight that the Earth receives.

Milanković calculated that the Earth went through regular cycles of cooling and warming, resulting in glacial periods (when ice sheets on Earth cover greater areas) that recurred in cycles that were in step with the combined effects of eccentricity, obliquity and precession.

The problem was that the cycles didn't fit with the dates of the ice ages that geologists had identified in glacial rocks. Milanković's calculations produced many more glacial periods. Though his theory was rejected, Milanković died confident he had solved the mystery of the ice ages.

PRISONER OF WAR

Milutin Milanković (1879–1958) was born in a small village in what is now Croatia. He studied engineering in Vienna and began a career building bridges and dams. He was on his honeymoon visiting Dalj, his family's village, when World War I broke out. Ethnically Serbian, he was arrested and imprisoned by the Austro–Hungarian army. Fortunately, he was permitted to serve his sentence under house arrest in Budapest. He was also allowed to visit a nearby library. He spent the war working on the unsolved mystery of the ice ages.

AMATEUR THEORY

Guy Stewart Callendar was another amateur climate scientist who decided to revisit an abandoned climate theory. He wanted to prove Arrhenius's theory that the land temperature was rising due to increased carbon dioxide. To do this he compiled temperature records that had been collected at weather stations around the world over a 50-year period.

Callendar presented his theory to members of the Royal Meteorological Society in London in 1938. He believed the Earth was getting warmer and that carbon dioxide from burning fossil fuels was responsible for causing this rise in temperature. He thought there could be a one-degree rise in a couple of centuries. Meteorologists still thought it was inconceivable that the tiny amount of carbon dioxide in our atmosphere could alter climate.

PART-TIME METEOROLOGIST

Guy Stewart Callendar (1898–1964) was an English engineer. Like his predecessors in early climate science, his interest in climate was a hobby. In his spare time, he collected weather information from weather stations around the world. Blind in one eye (thanks to an accident involving his brother and a pin), he was unfit for active service in both world wars. He worked for the war effort during World War II, but chose not to develop anything destructive; instead, he helped devise an airfield fog dispersal system called FIDO (Fog Investigation and Dispersal Operation) that saved the lives of many fighter pilots.

RISING TEMPERATURE

American Charles David Keeling (1928–2005) wanted to know exactly how much carbon dioxide there was in the atmosphere. The first thing he had to do was to build some equipment to do the job accurately. He set up his instruments at a research station on Mauna Loa, one of the five volcanoes that make up the main island of Hawaii.

Keeling's graph also showed that each spring and summer, as plants take in carbon dioxide during the growing season, the amount of carbon dioxide in the atmosphere falls slightly. In winter, carbon dioxide levels increase as the plants die.

In 1958, his equipment began taking readings every hour. For the first time, the level of carbon dioxide in the atmosphere was being systematically and accurately recorded.

Two years later, Keeling produced a graph that showed the level of carbon dioxide in Earth's atmosphere was indeed rising.

BOLD PREDICTION

In 1959, a young man wearing a bow tie spoke to a meeting of the National Academy of Science in Washington. He was Swedish meteorologist Bert Bolin, who was interested in the carbon in oceans and in living things, as well as the carbon in the atmosphere. He surprised his audience by making the bold prediction that the amount of carbon dioxide in the atmosphere would increase by 25–30 per cent above the level it had been before the Industrial Revolution, not in thousands of years, not in hundreds of years, but by the end of the century – just 40 years away. He also said it would be caused by humans burning fossil fuels. Bolin's daring statement was mentioned in *The New York Times*. The journalist said that "the effect on climate allegedly might be radical". This was possibly the first time climate change had made the newspapers.

WINDOWS INTO THE PAST

Before they could confidently predict what Earth's future climate would be like, scientists needed to learn about the patterns of Earth's climate long before the 130 years of meteorological records. Fortunately, past climates have left their mark on Earth. Our amazing planet has been collecting the information for us. During the mid-twentieth century, some climate scientists began exploring the world in search of information about past climates. This often took them to remote and inhospitable places.

Most people know about fossils – remains or imprints of plants and animals from millions of years ago that are preserved in rocks. They tell us which creatures were living on Earth in the distant past. But the Earth also stores information that can tell scientists what the atmosphere was like long ago. These natural archives are known as climate proxies. A proxy is something that acts as a substitute for something else. In this case, samples of these natural things are a substitute for meteorological records.

THE EARTH'S ARCHIVES

Trees can be preserved for thousands of years if they happen to fall into an anaerobic bog or become frozen in a glacier. The tree rings in these ancient trees can tell us about the weather in past years. Trees don't grow as much in harsh weather conditions, so the rings are thinner in colder years.

Coral forms bands similar to tree rings as it grows. Living coral reefs can be up to 10,000 years old. Ancient dead corals survive in some places, stranded on land after a fall in sea levels. Corals are made from calcium carbonate and the oxygen atoms within them preserve information from past times.

Stalactites and stalagmites form when mineral-rich groundwater drips from a cave roof. They grow at a rate of about 0.13 millimetres per year and their age can be determined using radiometric dating. Analysing them can tell us about temperature and rainfall in the past. From the stalagmites in Hulu Cave, scientists learned about monsoons thousands of years ago.

Pollen from flowers can survive for many thousands of years. Ancient pollen grains are found in lake beds, and they can be radiocarbon dated. Pollen from different plants has a unique structure and scientists can identify which plants were growing long ago. They can then work out what the climate would have been like when the pollen was deposited.

DATING

Rocks and other natural formations, such as stalactites, contain tiny amounts of radioactive elements that have survived from when they formed. These elements decrease or "decay" over time and turn into different elements. Radiometric dating measures the amount of decay, and from that the age of the rock can be estimated. In dating stalactites, the decay of uranium into thorium is measured.

Carbon-14 is a weakly radioactive isotope of carbon that exists in the atmosphere. Some carbon-14 finds its way into plants and animals. When they die, the amount of carbon-14 decreases as the plants and animals decompose. Scientists can measure the amount of carbon-14 in human and animal remains (such as bones, teeth and hair) that archeologists dig up. They can also test plant remains, such as grain seeds. They can then estimate the date when the person, animal or plant died. Artefacts made from natural substances, such as wool, bone and wood, can also be dated.

ICE CORES

The coldest areas of the Earth, the polar regions, are covered with ice that holds information about past climates. Each year snow falls at the poles and remains frozen due to the sub-zero temperatures, and is compressed under its own weight until it becomes ice. This makes the layer of ice thicker and thicker. Information about past climates is trapped in the ice.

The ice sheet on Greenland is up to three kilometres thick. In Antarctica, there are places where the ice is more than four kilometres thick.

Scientists bore down with a hollow drill and pull out cores of ice about ten centimetres in diameter and two to six metres long. The drill then burrows down further and another ice core is extracted. The process is repeated, providing a continuous sample of ice. It took two decades to develop the technique to drill deep and retrieve long ice cores without disturbing the layers of ice or damaging the bubbles of ancient air they contain.

Miraculously, climate scientists can distinguish within the cores layers of ice created in individual years. They can count back through the layers and examine the condition of the ice at specific times. The ice core record goes back hundreds of thousands of years.

Beginning in the 1960s, ice core drilling sites were set up in Greenland and Antarctica. Over the years techniques improved, drills burrowed deeper and the ice cores extracted grew longer and longer. The deeper into the ice the scientists drilled, the older the ice. The longest ice cores are more than three kilometres long and can take many months to collect.

At Vostok Station on Antarctica, between the 1970s and 1980s, a Russian and French team extracted an ice core 2083 metres long. This core contained layers of ice that dated back 160,000 years.

VOSTOK STATION

A major Russian Antarctic research station, Vostok, began operation in 1957. It is situated near the South Geomagnetic Pole. Vostok is the coldest place on Earth. The lowest temperature ever was recorded there on 21 July 1983. It was -89.2°C. The highest temperature recorded at Vostok is -12.2°C! The station is 3448 metres above sea level. Supplies are delivered once a year by snow vehicles that crawl 1260 kilometres across the ice from the nearest sea coast.

SECRETS IN THE ICE

Thin layers of ash in an ice core tell scientists that there was a major volcanic eruption or a huge forest fire somewhere on Earth. A layer of dust can indicate that there was a large area of desert at a particular time. Ice cores also contain tiny bubbles of ancient air. Scientists can extract the air and measure the levels of carbon dioxide and methane. This is not an estimate. It is a precise measurement of the concentration of these gases in the air preserved from hundreds of thousands of years ago.

Oxygen atoms exist in different forms known as isotopes, each with a slightly different molecular structure. When studying ice cores, scientists can determine the ratio of the different oxygen isotopes in the water molecules in each layer of ice. From this they are able to estimate the temperature of the air at the time the layer of ice was formed.

A one-metre-long section of Antarctic ice core, with a clearly visible layer of ash from a volcano.

DEEP SEA

Scientists have found the oldest record of the Earth's climate beneath the seas. In the deepest parts of the oceans, sediment has been accumulating for millions of years. Scientists carefully extract cores of the soft muddy sediment, which is made up almost entirely of fossils of tiny plants and animals. These ocean cores aren't as long as ice cores, but they are more compressed. Each millimetre or two represents a century.

The most predominant remains in the sediment are the tiny fossilised shells of microscopic animals called Foraminifera (usually shortened to forams). The minute creatures that occupied the shells have lived and died in Earth's oceans for millions of years. When the forams die, their shells sink to the ocean floor. They hold within them a record of the chemistry of the sea water when they were alive.

The oxygen in the molecules of sea water finds its way into the calcium carbonate molecules that make up foram shells (and corals). These tiny shells hold the secret of how much ice there was on Earth when they were alive. This enables scientists to reconstruct the climate millions of years ago.

GOING BACK IN TIME: WHAT WE LEARN FROM CLIMATE PROXIES

TYPE OF PROXY	HOW FAR BACK?	WHAT DO WE LEARN ABOUT?
TREE RINGS	10,000 years	temperate forests
CORALS	125,000 years	tropical regions
STALACTITES AND STALAGMITES	500,000 years	rainfall
ICE CORES	800,000 years	polar regions
LAKE SEDIMENTS	2 million years	land vegetation
DEEP SEA SEDIMENTS	55 million years	deep oceans, polar ice volume

MYSTERY SOLVED

With the aid of climate proxies scientists have constructed a climate history that goes back millions of years. They charted the rise and fall in temperature calculated from oxygen isotopes in ancient ice, corals and forams. They also plotted the levels of carbon dioxide from the air bubbles in ice cores. The two graphs are strikingly similar, indicating a strong link between Earth's temperature and the level of carbon dioxide in the atmosphere.

What's more, the ocean sediment cores showed that within the last ice age, there were alternating cold periods and warmer periods. And these matched Milanković's calculations from 50 years earlier. The cores provided evidence that the variations in Earth's orbit in relation to the Sun did indeed bring about conditions that triggered an ice age.

The mystery of the ice ages had finally been solved. We

now know that within the Quaternary ice age (which began around 2.6 million years ago, and hasn't finished yet) there have been around 40 fluctuations in temperature. The colder periods are called glacials. They lasted between 70,000 and 90,000 years. The warmer periods were shorter – 10,000 to 30,000 years – and are called interglacials. (We're in an interglacial period now.)

> *"What climate history can tell us is what has happened and what can happen ... What it can't tell is what will happen under precisely the conditions we have now because we don't have any ice core samples with levels of CO_2 as high as they are today."*
> **Eric Wolff, leader of the British Antarctic Survey's Chemistry and Past Climate team, 2010**

Our understanding of the orbital influences on our climate was finally becoming clear. But at the same time the rapid rise in carbon dioxide levels produced by human activities was beginning to overwhelm the impacts of the Milanković cycles.

CORRELATION BETWEEN TEMPERATURE AND CO_2 LEVELS

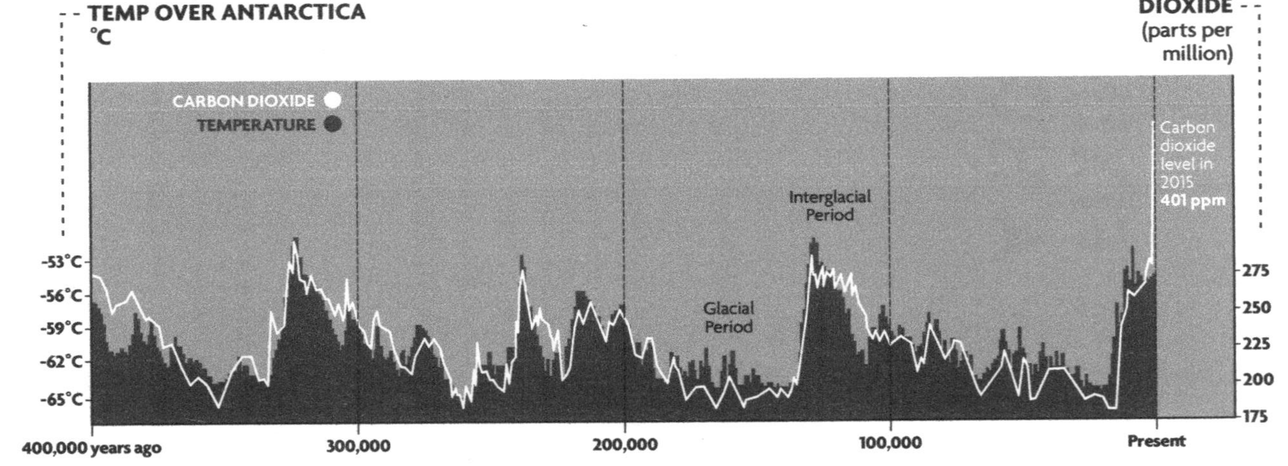

9

FIXING A HOLE

1964
Adelaide, Australia

It was hard at first leaving my friends behind in England, but my family has been in Australia for a year now and I'm glad we emigrated.

We bought a house. In England, we lived in a council house. This one is ours. It's very modern. We have a shower. I'd never even seen a shower before we came to Australia. The glass shower screen has a picture of a shark and a chest of sunken treasure etched on it. We even have a laundry!

We didn't have a fridge in England, just a marble shelf in the pantry where we kept the milk and cheese and meat. Now we have two! I love opening the fridge after school, and standing there deciding what to eat and drink with the cold air cooling me down. Dad bought the other fridge second-hand. We keep it in the garage just for soft drink and beer.

We have our first car too. A Holden FB, white with pink trim. Mum and Dad have both learned to drive. We go for a drive every weekend. My aunty and uncle and two cousins usually go too, in their Falcon. We drive for hours, just watching the countryside go by. Mum packs sandwiches and a thermos of tea so we can stop somewhere for a picnic. We use paper plates and plastic knives and spoons, which we can throw away so she doesn't have any washing up to do when we get home.

Winter smells different in Australia. There's no smog. It still gets cold enough to need heating – a gas fire in the lounge room and an electric heater for when I'm doing homework in my bedroom. Dad missed having a coal fire last winter, but Mum said she didn't miss having to clean up all the ashes.

Summer in Australia is great. If it's been a hot day, we go to the beach after dinner. You can drive your car onto the beach. We have icy poles almost every day and a tub of neapolitan ice-cream in the fridge all the time.

Dad loves the hot weather and wears shorts to work. He is very brown from spending so much time out in the garden or at the beach. I'm getting a tan too. When it's hot, my brother likes to play under the sprinkler out on the lawn. For Christmas, he wants one of those plastic water slides.

Mum isn't so keen on the hot weather. She says that next year we will get an air conditioner. She doesn't like the flies and the mosquitoes either. She's always going round the house with a can of flyspray. She sprays outside as well, to get rid of all the ants, especially the bull ants, and she's scared there will be redback spiders. Mum gets sunburned very easily and the mozzies love to bite her. She won't go outside in summer unless she's covered herself with sunburn cream and Aerogard.

It took me a while to get used to my new school. My classroom is a wooden portable, which gets very cold in winter. There is a wood stove in the corner and we have to take it in turns to be stove monitor and keep the fire alight. I've made some friends now and I go to their houses on the weekends. We listen to records and do each other's hair. Last weekend they put my hair up in a French roll. They used so much hairspray I thought I was going to choke.

I'm looking forward to the school holidays. We are going to drive to Narooma in New South Wales. In England, when we went on holidays, we had to catch the train. When we get home, I'll spend a lot of time at the beach with my friends improving my tan.

There's nothing that's old in this country. Everything is modern. There are drive-in movie theatres and motels and radio stations that play pop music all day long. It feels like we didn't just move to the other side of the world when we came to Australia. It's like we moved to the future.

Carole Ross, migrant, aged 14

During the 1960s, more and more people in the developed world were buying products that were not essential for daily life. Items for entertainment, such as televisions, were no longer considered to be luxuries. The trend towards disposable products was starting to take hold. Electricity and water were cheap, and no one worried about how much of either they used. Nobody knew that the electricity and gas they used in their homes, both produced by coal-fired processes, were steadily adding carbon dioxide to the atmosphere.

It has been estimated that in the 1970s, at any given time, there were more than 20 aerosol spray cans in use in an average US household.

People were completely unaware that man-made chemicals used in refrigeration and spray cans were accumulating in the atmosphere and had the potential to cause an environmental disaster in part of Earth's atmosphere.

DISCOVERY

A German–Swiss scientist called Christian Schönbein (1799–1868) reported in 1839 that gas produced during some of his experiments had a particular smell – the same sharp smell he had observed after thunderstorms. He called the gas ozone, from the Greek word for smell.

In 1913, French physicists Charles Fabry (1867–1945) and Henri Buisson (1873–1944) discovered that ozone was spread thinly throughout the atmosphere, with most of it concentrated in the lower stratosphere, 25–35 kilometres above us. This has become known as the ozone layer.

BAD NEARBY, GOOD UP HIGH

The ozone created by car exhaust fumes is a dangerous pollutant in the lower troposphere where we live. But in the upper atmosphere, ozone is crucial to our survival.

Ozone at ground level can cause chest pain, coughing and throat irritation. It can damage lungs and make bronchitis and asthma worse. However high above us in the stratosphere the ozone layer protects us, and all life on Earth, by absorbing or reflecting away some of the damaging ultraviolet radiation that comes from the Sun.

Ozone molecules consist of three atoms of oxygen (O_3). The oxygen that we breathe consists of molecules with two oxygen atoms (O_2). Only three in every 10 million molecules in the atmosphere are ozone molecules.

Ozone in the stratosphere is naturally occurring. It forms when ultraviolet light hits oxygen molecules, splitting them into two oxygen atoms. These single oxygen atoms then combine with other oxygen molecules and ozone is reformed. This is known as the oxygen–ozone cycle and it continued undisturbed for millions of years until certain chemicals were manufactured.

MIDGLEY RETURNS

In 1928, Thomas Midgley Jr was working for a division of General Motors called Frigidaire that made fridges. He was given the task of finding a replacement for ammonia and sulphur dioxide, which were the gases used in refrigeration. They were known to be corrosive, flammable and toxic to animals and plants. People had died from gas leaks and explosions caused by early refrigeration units. Midgley had the idea that a compound containing fluorine might work, and he synthesised something called dichlorodifluoromethane (CCl_2F_2). Frigidaire gave it the catchier brand name of Freon.

"Midgley ... had more impact on the atmosphere than any other single organism in Earth history."
J R McNeill, environmental historian, 2000

Midgley staged another of his public safety demonstrations by inhaling some of the gas and then blowing out a candle to show that it was neither toxic nor flammable.

Though Midgley knew full well that lead in petrol was extremely toxic, he was completely unaware of the impending global impact of his latest creation.

RECKLESS CHEMIST

Thomas Midgley Jr (1889–1944) was the son and grandson of inventors. His first job was in the Inventions Department of the National Cash Register Company. He then worked in his father's tyre factory, helping improve the design of tyre treads.

While working on his famous chemical discoveries,

he was reckless with his own health, and his family's safety, in pursuit of his goals. His wife and children suffered from tellurium poisoning during his search for an anti-knock agent. Thomas Midgley contracted polio when he was 51. Ever the inventor, he built a device to help him get from his bed to his wheelchair. A few years later, he became tangled in the cords of this contraption and was strangled to death. He died unaware that he would be remembered as the creator of two of the most harmful pollutants ever.

COOL

Other gases containing chlorine, fluorine and carbon atoms were synthesised. Collectively they are known as chlorofluorocarbons or CFCs for short. Use of these gases was soon widespread as Frigidaire expanded into the rapidly growing market for air conditioning and supermarket refrigeration.

CFCs are not naturally occurring. They are cheap and easy to make, also non-toxic and non-flammable. Manufacturers of CFCs believed they were very stable and wouldn't break down into other substances that were dangerous to humans. CFCs appeared to be completely safe replacements for the earlier coolants.

By the 1950s, other applications for CFCs had been

discovered, including their use in aerosol cans. Spray cans had become very popular and contained everything from insecticides, cleaning products and paint to toiletries, such as hairspray, shaving cream and deodorant. Every aerosol can needs something to make the product inside spray out in a fine, even stream. Chemicals called propellants are used for this. CFCs soon became the most popular propellant.

It was some time before anyone realised there was any connection between the discovery of the ozone layer and the creation of CFCs.

THE LINK

By the 1970s, the total production of CFCs around the world had reached 1 million tonnes per year. Measurements of atmospheric CFCs seemed to indicate that the manufacturers were right – they were stable. In fact, it looked like all the CFC molecules ever manufactured were still in the atmosphere. Scientists became concerned that they might stay intact long enough to end up in the stratosphere where they would finally be broken down by strong ultraviolet radiation, releasing chlorine atoms.

In 1973, chemists American F Sherwood Rowland and Mexican Mario Molina showed how CFCs had reached the ozone layer, where they did indeed finally break down. Chlorine can break apart ozone molecules. Each single chlorine atom contained in all those tonnes of CFCs, once released, had the capability of destroying thousands of ozone molecules.

Other man-made chemicals used in dry-cleaning, fire-extinguishers and pest extermination, most of them containing chlorine atoms, were also identified as having the ability to break down ozone. These chemicals were grouped together with CFCs and called ozone-depleting substances. But what effect would loss of stratospheric ozone have?

Molina and Rowland received the Nobel Prize in Chemistry in 1995 "for their work in atmospheric chemistry, particularly concerning the formation and decomposition of ozone".

NOBLE NOBEL LAUREATE

Mario Molina (1943–) was born in Mexico City. He had an early interest in science, and created a chemistry laboratory in a little-used bathroom in his family home. He was a good enough violinist to consider music as a career, but chose science instead. He moved to America in 1968. Aware that some scientific research was for harmful causes such as weapons development, he wanted to work in a field that would benefit mankind.

He joined the research team of Professor F Sherwood Rowland (1927–2012) in 1973. They chose a research topic that was far from both men's previous fields of experience – the fate of CFCs in the stratosphere. Molina was the first Mexican recipient of a Nobel Prize. He later returned to Mexico City and worked successfully on reducing air pollution there.

PEOPLE POWER

If there was less ozone in the atmosphere, it would mean more ultraviolet rays from the Sun would reach the surface of the Earth. More ultraviolet would be harmful to people's skin and eyes, resulting in an increase in skin cancer and cataracts (cloudy areas on the lens of the eye). Another negative effect of higher levels of ultraviolet is reduced productivity in food plants.

No one had proved that CFCs were breaking down and reducing ozone. It was still a theory. Manufacturers of CFCs, confident their products were safe, didn't want to stop making such a lucrative product. They tried to discredit scientific reports that CFCs affected ozone. They blamed volcanoes for extra chlorine in the stratosphere, and accused scientists of overreacting.

The issue had been reported in the media. Environmental and consumer groups encouraged people to stop buying products in spray cans that contained CFCs. If governments weren't going to do something about it, ordinary people were. They found alternatives – roll-on deodorants and other products in pump-action sprays. Sales of spray cans in the USA started to drop, even though there was still no proof that the ozone layer was being damaged.

But industries worldwide were continuing to come up with new uses for CFCs and production was increasing.

DISBELIEF

The British Antarctic Survey (BAS) had been measuring the ozone level above Antarctica since the 1950s. The general consensus was that since then ozone levels had decreased by a small amount, two to three per cent. In 1982, one of the BAS team, Joseph Farman, was studying the latest temperature measurements from Antarctica and found that there had been a sudden drop in ozone levels. During late winter and early spring the ozone layer above the Antarctic had shrunk by 30–50 per cent. He thought the fall in ozone levels was so dramatic it couldn't be accurate.

> *"Everyone knows the hole in the ozone layer is there, but it should horrify them. You know, this is something which man did in 15 years. And one simply has to say: if you invent something, let's take it slowly until we're reasonably satisfied we can't see how it can be dangerous."*
> **Joseph Farman, geophysicist, 1999**

Farman had good reason to doubt the results. Above the Earth, a well-equipped NASA satellite was circling the globe taking 140,000 atmospheric readings a day. Since this high-tech equipment hadn't picked up a change in ozone levels, Farman thought

BAS's ageing equipment must be faulty. The British team acquired new equipment and continued monitoring Antarctic ozone levels for another three years before they were convinced that ozone levels really were falling.

NASA staff had noticed the low ozone levels over Antarctica in 1983. They didn't believe their readings either. They also assumed there must have been an equipment problem. The NASA satellite was collecting so much information that there was a five-year backlog of data waiting to be carefully studied by scientists.

CAUTIOUS GEOPHYSICIST

Joseph Farman (1930–2013) was one of the British Antarctic Survey team that discovered the hole in the ozone layer.
He was employed by an aircraft company developing guided missiles when he saw a job advertisement in a newspaper seeking scientists to take part in Antarctic research. He applied, was accepted and spent a number of years working in Antarctica. While Molina and Sherwood won the Nobel Prize for their work on ozone, Farman had to be content with several medals. After he retired from the British Antarctic Survey, he worked in the Cambridge University Chemistry Department, cycling there daily until the day before his death.

OZONE HOLE

Finally, in 1985, the BAS team reported that the amount of ozone in the atmosphere above Antarctica was thinning. This depletion of ozone became known as the hole in the ozone layer. CFCs breaking down in the stratosphere were to blame for most of the ozone depletion. The atmosphere above Antarctica experienced more ozone depletion than anywhere else because the chemical reactions that cause the breakdown of CFCs happen faster at lower air temperatures. Ozone destruction was at its highest in the Antarctic spring and summer, when there were long days and a lot of sunshine, but polar winds kept the air cold.

On Antarctica the summer maximum temperature is around -20°C, but scientists working there have to use SPF 30 sunscreen to avoid sunburn because of the high level of ultraviolet.

The ozone hole extended and affected inhabited areas beyond Antarctica including Australia and New Zealand. The "hole" was actually an area where the layer of ozone was thinner by as much as 65 per cent. There was ozone thinning above the Arctic as well, but it wasn't as severe as above Antarctica.

Ozone depletion was no longer a theory, it was proven. What could be done to stop the thinning of the ozone layer?

BAN

The USA had taken the first step to stop ozone depletion even before it was proved, banning the use of CFCs for non-essential purposes in 1978. The Scandinavian countries, Canada and others followed this example. But the European Community initially refused to implement bans.

Though there was an urgent need for world governments to cooperate in banning CFCs, there was disagreement and deadlock for almost a decade. But the public had turned against products that contained ozone depletors. Sales of spray cans had now fallen by 75 per cent. Even before bans were put in place, the manufacturers had begun to search for alternatives to CFCs. Once the existence of the ozone hole was confirmed, opposition to bans disappeared and new negotiations began.

In 1987, international agreement was reached that required wealthy industrialised countries to reduce manufacture of the most widely used CFCs by 50 per cent by the year 2000. Poorer developing countries were given ten years longer. This agreement is known as the Montreal Protocol on Substances that Deplete the Ozone Layer. In 2009 it became the first international environment treaty to be agreed to by all participants.

PHASE-OUT

Since 1987, the Montreal Protocol has been expanded and strengthened. Other ozone-depleting chemicals have been added to the list to be phased out. The dates for phase-outs have been brought forwards.

This unprecedented international cooperation, guided by scientific theory and discovery, has led to the elimination of a number of industrial chemicals that were once produced in millions of tonnes.

Australia, because it is so close to Antarctica, was one of the countries that stood to suffer most from ozone depletion. The Australian government has therefore always been a strong supporter of control of ozone-depleting substances.

The maximum size of the ozone hole was reached in 2006 when it was 27.5 million square kilometres. That's three and a half times the size of Australia.

In 2000, the level of CFCs in the atmosphere peaked and has been slowly falling ever since. The USA and Europe ended manufacture of CFCs in 1996, and by 2010 so did the rest of the world. It is predicted that by 2050 the hole in the ozone layer should be completely healed.

POSTSCRIPT

New evidence was released in 2014 that shows that the job of ridding the world of ozone-depleting substances isn't over yet. A British team measuring air trapped in snow in Greenland and in Tasmania discovered that concentrations of four banned ozone-depleting substances are still rising. One in particular, CFC-113a, is rising quickly. There are other ozone depletors that haven't been detected previously. Though the amounts are small, it is a worrying trend.

Carbon tetrachloride is another banned ozone-depleting substance that is not disappearing from the atmosphere as fast as expected. No one knows why. Could someone in the world be illegally manufacturing these chemicals?

The international ban on ozone-depleting substances has been the most successful exercise that humans have undertaken to stop pollution of Earth's atmosphere by harmful chemicals that affect life on Earth. One problem in the atmosphere was on the way to being solved. But a much bigger one – global warming – was looming.

Scientists didn't know it yet, but CFCs weren't only ozone depleters – they were greenhouse gases as well.

10

HOTTING UP

21 August 1981

229 West 43rd Street, New York, USA

On Monday, I was on top of the world. It was my first day as a cadet journalist, and with The New York Times *no less! I thought it was my dream come true. Then the editor told me I'd be working with the science writer. Don't get me wrong, I know that science is important, but I'd been hoping for crime. I'd pictured myself staking out sleazy bars in a battered Dodge, eating burgers and drinking coffee, listening in to the police radio channel.*

I didn't do a very good job of hiding my disappointment.

"You should think yourself lucky, young lady," the editor said. "You'll be with Walter Sullivan, the top science writer in the country."

I tried hard to look enthusiastic.

"You'll learn from him, Lois."

Lois? It took me a while to realise he was referring to Lois Lane – you know, Superman's girlfriend.

Mr Sullivan was no Superman, not even a Clark Kent. He was sitting at his desk looking miserable and holding the right side of his face. He had a toothache. I was ready to start writing, to work till midnight to meet deadlines. But he gave me a bunch of stuff to read, mainly his old articles, while he sat and groaned.

There's no doubt about it, he has written about some cool things – black holes, the search for extraterrestrial intelligence, tectonic plates. He's won awards and medals too.

Two days later, and I'm still reading. The editor is getting antsy. He wants a science article today. Mr Sullivan is typing.

"So what are you writing about, Mr Sullivan?" I ask.

He slides a few xeroxed sheets towards me.

"Some of the guys out at Goddard have written this," he says, a little slurred. The right side of his face is swollen.

That's the first time I've heard him speak a full sentence.

"Goddard?" I say. "You mean the NASA Space Flight Centre?"

He grunts.

This is more like it. Space exploration is my kind of thing.

My excitement doesn't last long. The article has got nothing to do with space and other galaxies. It's an advance copy of a very dull scientific paper that will be published in a science magazine next week, about something called the greenhouse effect. The NASA guys have predicted that the world is going to be a couple of degrees warmer next century. Next century? Who cares? Doesn't that mean it won't be so cold in winter? Sounds good to me. I plough through the paper. There are a lot of scientific calculations that I don't understand.

Mr Sullivan gets me to take his article to a subeditor, who takes it to the editor. I wait.

The subeditor comes back and throws the article at me.

"He doesn't like it," he says, "He wants more spice."

I have to go back and break the news to Mr Sullivan.

"We could go out to Goddard and talk to these guys," I suggest. "It'll only take 20 minutes on the M."

Mr Sullivan starts to shake his head, then thinks better of it. One of the office girls has made him a dental appointment. He gets his jacket, even though it must be 80 degrees out there.

I think about going out to Goddard myself, but instead I read the scientific paper again. Now that I look at it a second time, there are some interesting bits. I underline them with a red pen.

Mr Sullivan's back. He can't talk at all because his mouth is numb. He's more cheerful though, now he's not in pain.

I decide to go for broke.

"I think I found some spice for the article," I say.

He's staring at the tops of the trees out in the street, smiling. I think the dentist must have given him gas.

"You should put in something about how seas could be rising and ten per cent of New Jersey will be under water."

He nods.

"And mention the dinosaurs."

I thought that would get his attention, but no, he just stares at the treetops.

I decide to have a go myself. See what I can do with it. Just for the experience. I pull the typewriter over, put in a clean sheet of paper and start retyping the piece. I add the prediction that the increase in temperature will mean that sea levels could rise by 15 feet, maybe even 30. I add that the last time that happened was long ago before the last ice age. It might end up as hot as in the age of the dinosaurs. I give it a better headline: STUDY FINDS WARMING TREND THAT COULD RAISE SEA LEVELS.

It was fun pretending to be a real journalist for a little while.

Then the subeditor comes in and snatches it out of the typewriter.

"But …"

Cassie Montgomery, cadet journalist, aged 17

During the 1980s, the accumulation of knowledge about how the Earth's climate works began to accelerate – as did the level of carbon dioxide in the atmosphere. Computer modelling was improving, giving climate scientists new confidence that our climate was changing. And compared to previous eras, the change was fast.

HEADLINES

Neither politicians nor the general public read the scientific journals where breakthroughs in climate science were being revealed. It was only when newspapers began reporting on the increasing urgency of climate change that the rest of the world started to take notice.

> *"The climate change induced by anthropogenic [human] release of CO_2 is likely to be the most fascinating global geophysical experiment that man will ever conduct."*
> **James Hansen, atmospheric physicist, and others, 1981**

There was an article in *The New York Times* in August 1981 about a study by atmospheric scientists at the NASA Institute for Space Studies led by Dr James Hansen. Using temperature records observed over the previous century, they found that global temperatures had risen 0.4°C. At the same time, the level of carbon dioxide in the atmosphere had risen from 280 to 340 ppm. Computer modelling predicted unprecedented temperature rises in the future. This, Hansen said, could result in the world's ice sheets melting. And that would cause sea levels to rise, and areas of land to be submerged. The only way to prevent this scenario was to stop burning fossil fuels.

The report made the front page.

ON LAND AND SEA

The most obvious way to gauge if the Earth is experiencing global warming is to compare the temperature from day to day, from year to year, in different places around the world. People had, of course, been measuring temperature since the invention of the thermometer, but, as James Hansen noted in his article, official records were limited. The measurements were mostly from the Northern Hemisphere, and mainly taken on land. There were few records taken on the 71 per cent of the Earth's surface that is ocean. The fundamental measurement of Earth's warming needed urgent work.

A team of British scientists followed up the NASA study, trying to fill in the gaps. They studied millions of meteorological observations from more than a thousand locations to produce an analysis of the average global temperature from 1851 to 1984. The team confirmed that there was a clear warming trend over the 133-year period. And the three warmest years had all come in the

Early thermometers were invented in the 1600s, but they were inaccurate. The mercury thermometer was invented by Daniel Gabriel Fahrenheit in 1714. It was the first thermometer to give accurate readings.

previous five years – 1980, 1981 and 1983. The Earth was getting hotter.

JUST A TRACE

So far, carbon dioxide had been the focus of attention when it came to the greenhouse effect. That was about to change. Human activities were contributing to the rising levels not only of carbon dioxide, but other greenhouse gases. The amount of carbon dioxide in the atmosphere was so small it was hard for people to believe it had the ability to alter climate. The amounts of the other trace gases were even smaller.

The Intergovernmental Panel on Climate Change lists 88 greenhouse gases other than carbon dioxide.

BAD NEWS

Not long after it became known that CFCs depleted ozone in the atmosphere, there was more bad news. In 1975, an Indian scientist called Veerabhadran Ramanathan was shocked to discover that CFCs absorbed infra-red radiation. That meant CFCs didn't only destroy the ozone layer, they were also greenhouse gases.

Ramanathan later conducted a study of the other greenhouse gases. He found that some were strong absorbers of heat in the atmosphere, and some had long atmospheric lifetimes.

GREENHOUSE GAS GURU

Veerabhadran Ramanathan (1944–) was born in Madurai, India. He began his working life as an engineer for a refrigeration company, but moved to the USA to study. He was looking forward to driving a big, fast car. He began studying atmospheric science, and in particular the atmosphere of Venus. He learned about global warming and changed his mind about fast cars, and decided to concentrate on the atmosphere of Earth. From his early work in refrigeration, he had personal knowledge of CFCs leaking into the atmosphere. Ramanathan was elected to the Pontifical Academy of Sciences at the Vatican by Pope John Paul II in 2004. He has met with other religious leaders, such as the Dalai Lama, asking them to encourage their followers to take action on climate change.

SYNTHETIC GREENHOUSE GASES

CFCs are used in a wide range of industrial processes, such as aluminium manufacture, refrigeration and cleaning. They make up just one percent of our greenhouse gas emissions, but they are potent. They survive in the atmosphere for up to 1700 years and have global warming potentials of up to 14,400. Hopefully, under the Montreal Protocol we no longer have to worry about them.

The gases manufactured to replace CFCs were hydrochlorofluorocarbons (HCFCs). Unfortunately they are also greenhouse gases with GWPs of up to 2300. They are still in production, but will eventually be phased out under the Montreal Protocol.

Hydrofluorocarbons (HFCs) were next to be used as a replacement for CFCs. They do not deplete ozone, but they are strong greenhouse gases. There is currently no control over their production.

There are many other synthetic greenhouse gases, some of which survive in the atmosphere for up to 3000 years. Their GWPs range from 1 to 23,000.

METHANE

Ramanathan's work on greenhouse gases was more than 100 years after John Tyndall had experimented with the gas from his Bunsen burner and discovered that it absorbed heat. Methane is a powerful greenhouse gas with a GWP 28 times that of carbon dioxide, and yet it makes up only 0.0001 per cent of the atmosphere.

Whether methane comes from a natural or man-made source, it is always a particular type of microorganism called methanogens that actually produce the gas.

But scientists now knew how much of each trace gas had been in the atmosphere in the past, thanks to the air bubbles trapped in ice cores. They found that methane had been steadily increasing for

Cows produce around 200 litres of methane every day. There is some in their farts, but it is mainly from burping. Cattle contribute about ten per cent of Australia's methane emissions. Farmers are experimenting with different types of cattle feed (such as canola meal and brewers' waste) to reduce the amount of methane cattle produce.

the last 400 years. Modern measurements from the atmosphere showed that in recent times methane levels had risen rapidly – 11 per cent in the decade up to 1987.

Natural sources, mainly wetlands, are the largest source of methane emissions. However, humans have been contributing to the methane in the atmosphere by creating rice paddies, burning fuels (fossil fuels and biomass such as wood) and dumping food waste in rubbish tips. Cattle farming is also a source of methane, which is produced by cows during digestion.

Methane is the second most prevalent greenhouse gas, after carbon dioxide.

Termites are a natural source of methane, which is a by-product of their digestive process. However, by increasing the amount of grasslands through deforestation, we are providing more habitats for termites, which causes them to breed more. Scientists have estimated that termites could be producing up to 15 per cent of the world's methane emissions.

NITROUS OXIDE

Only eight per cent of the world's greenhouse gas emissions is made up of nitrous oxide, but it stays in the atmosphere for a long time. It has 265 times the global warming potential of carbon dioxide. It occurs naturally, mainly emitted from soil as part of the nitrogen cycle, which involves plants, animals and the atmosphere. The Earth has natural ways of removing this gas from the atmosphere. It is absorbed by bacteria and broken down by ultraviolet radiation. However, around 40 per cent of nitrous oxide in our atmosphere comes from human sources. Most comes from agricultural use of synthetic fertiliser (ammonium nitrate) rather than natural fertilisers, such as compost or manure.

The greenhouse gas with the highest GWP is sulphur hexafluoride (SF_6). Its GWP is 23,500 times that of carbon dioxide. SF_6 is used in the electrical industry as an insulator in high-voltage switches and in circuit-breakers. Between 1987 and 2006, it was used in Nike Air Max sports shoes that used pockets of this gas for cushioning.

FORECASTING THE FUTURE

Meteorologists had started to use computers to predict the weather in the 1950s, but those computers were primitive machines that took up a lot of space, and had little computing power. Gradually computers grew in power and shrank in size. At the same time, computer operators' skills improved. The predictions produced are called computer models.

A decade later, scientists created computer models, not just using information about the current state of the atmosphere, but also including estimates of future changes, such as increased carbon dioxide levels. They moved from forecasting the weather to forecasting climate change.

By the 1980s, a large amount of data was being collected by weather stations on land, by ships at sea and by planes and satellites in the atmosphere. This information could be fed directly into computers.

The World Meteorological Organization (WMO) now has a network of more than 11,000 weather stations around the world measuring land, air and sea temperatures. The Cape Grim Baseline Air Pollution Station, on the north-west tip of Tasmania, is one of 30 WMO Global Atmosphere Watch stations collecting information about the temperature and chemical composition of the air to monitor changes.

The Baseline Air Pollution Station
at Cape Grim, Tasmania.

DISCOVERING THE PAST

Thanks to Charles Keeling, evidence of carbon dioxide accumulating in the atmosphere was being collected at Mauna Loa in Hawaii, but by 1990 there were only 30 years of these records. To learn about carbon dioxide levels in the more distant past, data was needed from before Keeling began his measurements. The answer was once again trapped in ice.

In 2004, a three-kilometre long ice core from an Antarctic station at Dome C was extracted by the European Project for Ice Coring in Antarctica (EPICA). This multinational collaboration of ten countries revealed climate conditions up to 800,000 years ago. It provided hard evidence that present-day levels of greenhouse gases are higher than anything that the world has experienced in the last 800,000 years.

Climate scientists from 22 different countries are currently working together to find a site in Antarctica where the ice is deep enough to take them back 1.3 million years.

CONFERRING

In the 1800s, there had been just a few men, often amateurs, involved in climate research. They dabbled in climate science in their spare time when they weren't working on their main scientific enterprises. By the end of the twentieth century, there were thousands of professional scientists worldwide specialising in some aspect of climate science, many of them in institutions with departments that focused entirely on climate change. As evidence for climate change grew, so did the confidence of climate scientists. They made bolder predictions about how climate would change. It was time for the world to take action.

Several important international climate conferences took place during the 1980s. Because of growing knowledge of climate change issues, these conferences were no longer only of interest to meteorologists. Politicians, government advisors and representatives of industry now attended them. Newspapers around the world reported on the conferences.

SCIENTISTS SPEAK OUT

One of these conferences (with the clumsy title of Assessment of the Role of Carbon Dioxide and of Other Greenhouse Gases in Climate Variations and Associated Impacts) was held in Villach, Austria in 1985. It was a joint conference focusing on a report compiled by Bert Bolin (the Swedish meteorologist who in 1959 had made the claim that carbon dioxide levels were rising). He was making more strong statements – the amount of carbon dioxide would be double pre-industrial levels by 2030, the temperature would rise by 1.5–4.5°C and melting polar ice would mean that sea levels could rise up to 140 centimetres.

The onset of climate change had been brought forward again. Most scientists could see that climate change wasn't a distant threat, but one that could happen in decades – in their own lifetimes, or at least their children's. They were confident that the science of climate change was "solid". They confirmed that human activities were responsible for the rise in greenhouse gas levels, and the rise in temperature.

For the first time, climate scientists were bold enough to call for urgent action. They wanted governments around the world to set targets to reduce greenhouse gas emissions, just as they had with CFCs. At the World Conference on the Changing Atmosphere in Toronto, Canada in 1988, climate scientists said that carbon dioxide emissions must be reduced by 20 per cent by 2005.

CHAIRMAN BERT

There was a need for an international organisation that could assess climate research from all around the world, and bring governments together to take action. In 1988, the United Nations created the Intergovernmental Panel on Climate Change (IPCC). Its mission is to evaluate scientific research on human-induced climate change and publish regular reports so that governments can assess the risks and decide what action to take. There are 195 countries and more than 2000 scientists participating in the process.

Bert Bolin's prediction that carbon dioxide in the atmosphere would increase by around 30 per cent by the end of the century was well on the way to being proved correct. (In 2000, it reached 32 per cent above pre-industrial levels). Since then, Bolin had continued to work in climate change research. He was appointed as the IPCC's first chairman.

The first thing the IPCC had to do was prepare a report on the current state of the atmosphere, assess what impacts on climate there would be in the future, and then come up with strategies to limit them. It took the IPCC two years to produce their first report. When it was finally published, it said that it was "certain" that human greenhouse gas emissions were enhancing the greenhouse effect. The IPCC was more cautious about the effects, predicting a rise in average temperature of 1°C by 2025.

SCIENTIFIC STATESMAN

Swedish meteorologist Bert Bolin (1925–2007) was always interested in weather prediction, recording the temperature as a child. In 1950, he spent a year at the Institute for Advanced Study in the USA working with a team using the first ever computer to create a computerised weather forecast. In 1959, he was among the first scientists to predict that CO_2 levels were rising and the effect could be serious. In 1997, he was the first to say we had to keep emissions to a safe level. A modest and quietly spoken man, he was an outstanding communicator and diplomat, chairing many committees. He chaired the IPCC from 1988 to 1997.

11

ACTION

2 March 2007

United Nations General Assembly Hall, New York, USA

I have never been in a room that is so big, so high and so strange. It is like a spaceship made of wood and gold. The round ceiling is dark and dotted with points of bright light and looks like a window with a view of the stars. Except it is daytime. Rows and rows of comfortable leather seats face the podium. It is not like a theatre or a cinema, because in front of each row of seats there are desks, and positioned along the front of the desks are the names of all the 193 countries that are members of the United Nations.

"This is one of the most important rooms in the world," one of the UN guides told us when we arrived yesterday. "It's where leaders, from countries big and small, rich and poor, come together to bring peace and security to everyone on the planet."

But today, though the hall is full, there are no government leaders. Instead, seated at the desks, there are students from around the globe, representing those same countries. I am sitting behind the sign for my country, Ethiopia, and it fills me with pride.

The UN International School organised this conference and the topic is global warming. Many of the students here are children of diplomats living in New York. They go to the school, but I won an essay competition to be a representative of my country. It is the first time I have left my home.

The UN Secretary-General, Ban Ki-moon, spoke to us. He told us it was the first speech he had made in his new job. There have been lots of people giving talks, many of them Americans. The one who is speaking now is a scientist who works in America, but who comes from India. He has told us how global warming will affect everyone in the world, rich and poor, and how we all contribute to global warming, whether we know it or not. I thought rich people were to blame, like the people here in New York, with their fast cars, air conditioners and tall buildings with lights blazing day and night. He says that global warming is also caused by the clouds of pollution that hang over Asia, which are formed from smoke and soot produced by all the small fires that poor people light to cook their food. We are all at fault.

The Indian scientist speaks of his grandmother who was one of those women who lit fires for cooking in India. There are many women like that in Ethiopia. Every day, they build small fires from wood and dried dung. Each one crouches over her fire, smoke in her eyes, constantly fanning the flames to keep it alight long enough to cook a simple meal. The scientist says there are millions of people who cook like this around the world, and they produce greenhouse gases.

The girl sitting in front of me is crying. It is such a sad and hopeless story this man is telling us. There are tears in my eyes as well. I feel sad for the world too, but I also feel angry. Those poor women have no choice. They have no other way of cooking, no gleaming electric stoves like Americans have, that only require you to turn a knob to provide heat for as long as you want.

The scientist invites us to ask questions and there are many students who put up their hands. I am too shy to speak into a microphone, but now, as everyone is leaving the hall to go to lunch, I make my way to the front. I go to the Indian scientist. I do have a question that I want to ask him.

He is a kind man; I can see that in his eyes. He smiles at me and says he will be happy to answer my question.

"You made us cry with your talk about how we are changing our climate and making it worse," I say. "Can you tell me what you yourself are doing to help solve the problem?"

His smile fades, he looks at me. He says nothing.

Unnamed, student, aged 16

The scientist who spoke at the 2007 UN International School Conference "Global Warming: Confronting the Crisis" was Veerabhadran Ramanathan. Although he had made a huge contribution to the world's knowledge of greenhouse gases, this young African girl's question took him by surprise. He had devoted his career to the study of climate change, but he realised he had done little to reduce emissions in his personal life.

"All of my papers were like obituaries talking about how we are damaging the atmosphere ... This interaction [with the African student] made me commit to seeking solutions for the problems that were being uncovered by my science."
Veerabhadran Ramanathan, atmospheric scientist, 2014

BACKLASH

Back in the 1990s, many scientists, government leaders – even conservative UK Prime Minister Margaret Thatcher – as well as a growing number of ordinary people were concerned about climate change. It seemed that everyone was ready to take action to stop global warming – everyone except the rich and powerful industries that relied on fossil fuels, such as coalmining companies, electricity generators, car manufacturers, chemical corporations and the oil industry.

Just before the first report of the IPCC came out, a coalition of companies (including Exxon, British Petroleum and General Motors) banded together to form the Global Climate Coalition.

The aim of the Global Climate Coalition was not, as

the name of the group seemed to imply, to support climate action. Instead it claimed climate change didn't exist. To counter the statements made by climate scientists, they employed their own experts – people with distinguished backgrounds in physics, space research and weapons development, but not in climate science. They published articles casting doubt on climate science and claiming there was no global warming. The members of the Global Climate Coalition used their high profiles and large bank accounts to lobby governments not to pass laws to force industry to limit greenhouse gas emissions.

Other institutions and think tanks with similar aims to the Global Climate Coalition were formed. These powerful climate deniers claimed that there wasn't enough proof that climate change was happening, and the current rises in global temperature could be a short-term fluctuation. Millions of dollars were spent on advertising campaigns. They circulated videos to journalists and politicians that stated that increasing carbon dioxide levels was a good thing. Like Arhennius, 100 years earlier, they said it would increase crop growth and help feed the world's hungry.

KYOTO PROTOCOL

In its second report in 1995, the IPCC made firmer statements on global warming. They said that by about 2050 the average temperature would rise between 1.5 and 4.5°C.

The public sat back and waited for governments to take charge of reducing greenhouse emissions the way they had with CFCs. But some governments had been swayed by the arguments of the climate deniers. When world leaders met to draw up an agreement, it quickly became obvious they weren't going to repeat the firm actions of the Montreal Protocol.

The UN decided to exempt developing nations from any emissions targets. A number of countries objected to this. Australia was one of them. Our government said it would have "a devastating impact" on Australian industry.

Negotiations continued for two years. Finally at a conference in Japan, governments agreed on a target for greenhouse gas emissions reduction of just 5%. The agreement became known as the Kyoto Protocol. Fifty-five countries were required to sign up to the protocol before it could go ahead. Even with such a low target for reduction, the US Senate voted unanimously against supporting any IPCC treaty that did not include the developing nations. New Zealand signed the protocol, but the Australian government followed the US example and refused to sign.

Australia finally ratified the Kyoto Protocol in 2007. New Zealand pulled out in 2012.

FREE ENTERPRISE

Opposition to taking action on climate change came from those who were afraid they would lose profits if they took steps to reduce their carbon footprint. They believed in free enterprise, where governments have no power to hamper their business activities by imposing rules and regulations. They claimed there wasn't enough proof that climate change was a real and urgent threat.

This wasn't the first (or the last) time that industries had resisted taking action when their products or manufacturing methods proved to be bad for the health of people or the Earth. As we have seen, in the nineteenth century, factory owners and coal-fired electricity generators fought against laws that limited the amount of air pollution they produced. The company that added lead to petrol claimed there were no health risks, despite the fact their employees were dying from its effects. It was belief in free enterprise that led the tobacco industry to resist government regulations on tobacco sales, and to deny scientific claims that it was harmful to smokers' health. Manufacturers of asbestos also challenged claims that their product was harmful, dragging out court cases so that victims died before a verdict was reached. The CFC industry initially resisted finding safer alternatives. In 2004, car manufacturers sued the state of California over their Clean Cars Program, which at the time required car emissions to be cut by 30 per cent by 2016.

The principles of free enterprise put profits before the health of people and the health of the planet. There are still industries doing this today.

DENIER

Frederick Seitz (1911–2008) was an eminent and awarded American physicist. During World War II he was involved in developing the nuclear bomb. When he retired, he became a consultant for a major tobacco company, circulating claims that smoking was not bad for people's health, and attacking the US Environmental Protection Authority (EPA) and Surgeon General over their statements about the risks of inhaling second-hand smoke. Seitz was co-founder of the George C Marshall Institute, a think tank with the aim of spreading doubt about climate science. He claimed there was no proof that global warming was caused by human activity, and no need to limit greenhouse gas emissions. At his death he was remembered as a leader of climate change denial.

THE CURVE CONTINUES

Meanwhile the equipment at Mauna Loa Observatory on Hawaii was still recording the Earth's temperature. The resultant graph, now known as the Keeling Curve, continued its upwards climb.

KEELING CURVE

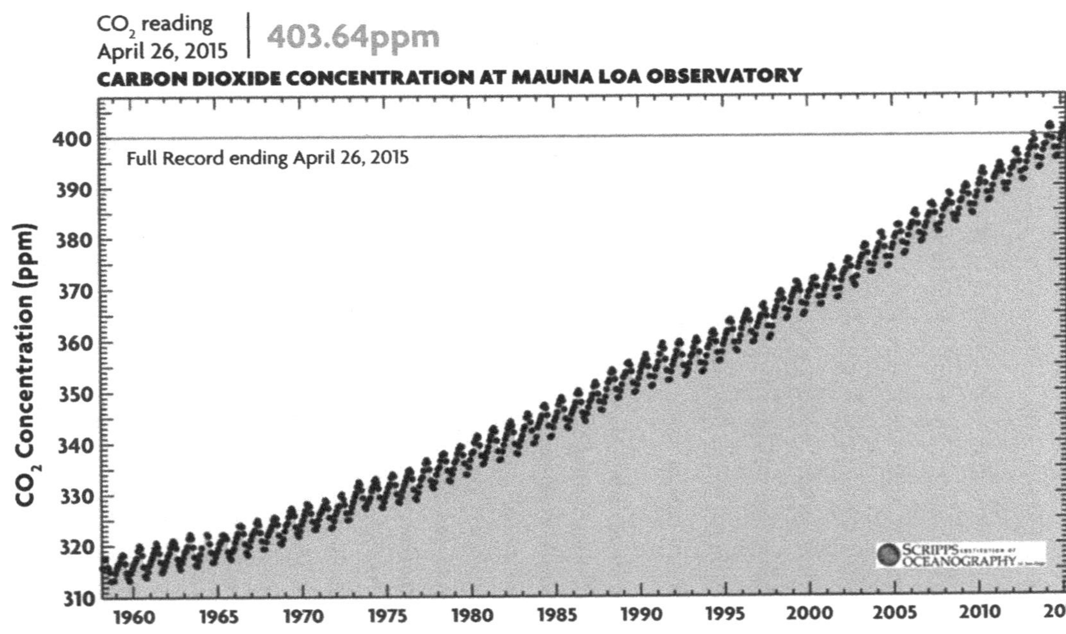

Despite this clear evidence that the Earth was warming, most industries took no notice. There were some, however, who realised there were business opportunities in combatting climate change. In 1997, Toyota began manufacture of a hybrid car that used electricity as well as petrol to power it. More companies began making wind turbines and solar panels to produce electricity from renewable sources.

After the failure of the Kyoto Protocol to introduce a meaningful target for greenhouse gas reductions, once again ordinary people took action where their governments had failed. The public's concern about the environment began to focus specifically on the state of the atmosphere and climate change. Many people started to take steps to reduce their own carbon footprints.

In 2000, the Global Climate Coalition, which had formed to refute the existence of climate change, collapsed. Member companies decided it was better for business to embrace climate action.

A personal carbon footprint is the amount of carbon dioxide produced by an individual. It includes greenhouse gas emissions generated directly from activities such as heating, transport and using appliances, but should include indirect emissions produced when goods that the person uses are manufactured. Carbon footprints can also be calculated for households, schools, businesses, towns and countries.

GRASSROOTS

In the early 2000s, groups large and small formed to spread

A solar electricity generator in China.

information about the urgency of climate change, and to lobby governments to take action on reducing greenhouse gas emissions.

Some of the earliest groups were in Britain. The Campaign against Climate Change began with rallies calling for then-President of the USA, George W Bush, to sign the Kyoto Protocol.

When the Nobel Peace Prize in 2007 was awarded jointly to the IPCC and former US Vice-President Al Gore for work done to spread education on climate change, it seemed that global climate action was close at hand. At a local level, many climate action groups formed in Australian towns and cities to encourage the Federal Government to act.

In 2008, international environmental groups such as Greenpeace began protesting about lack of climate action. In the USA, a group called 350.org formed to reflect the need to keep carbon dioxide levels below 350 ppm.

But free enterprise groups were also actively influencing governments. The promise of global climate action is yet to be fulfilled.

YOUTHFUL ACTION

The Australian Youth Climate Coalition (AYCC) was formed in 2006. Unlike other climate action groups, AYCC is exclusively for young people.

"At the AYCC we believe that the best way to build this movement is to give young people the tools to make it happen. It's our future at stake, and it's our creativity and vision that will inspire those around us to act."
AYCC website, 2015

It enables young Australians to take climate action in their schools, universities and the communities where they live.

The AYCC now has more than 120,000 members who take part in climate change campaigns all over the country. They organise school summits so that the voices of students, usually ignored in the climate debate, can be heard.

"So often, young Australians have heard themselves labelled the 'Me Generation'. But the irony of that tag is now nowhere more obvious than through the self-interest on display from older generations of decision makers on climate change."
Amanda McKenzie, AYCC founder, 2010

INSPIRED BY NATURE

When she was a child, Amanda McKenzie was a nature lover, learning about the natural world on family camping trips. She became concerned about the environment as a teenager, and began volunteering for environmental groups. In 2006, while hiking among 400-year-old swamp gums in Tasmania's Styx Forest, Amanda was inspired to do something about climate change. Amanda founded the Australian Youth Climate Coalition that same year, bringing together other young people who wanted action on climate. She is currently the CEO of the Climate Council, an organisation that provides Australians with independent information on climate change.

TOP: Australian Youth Climate Coalition (AYCC) members taking part in the PowerShift march, Melbourne 2013; **BOTTOM:** AYCC "flash mob" trying to get decision-makers to aim higher on climate, Brisbane 2009.

ADDICTED TO OIL

As the IPCC and climate action groups continue to fight for a safe climate, the amount of oil the world uses continues to soar. Oil refineries turn crude oil into the fuels that we consume continuously – petrol, diesel, jet fuel, liquefied petroleum gas (LPG).

The richest countries in the world are those with oil reserves beneath their land. When we see on the TV news that the price of oil has risen, they're talking about crude oil that is sold by oil-rich countries and bought by countries that don't have oil.

But oil is currently not a stable commodity. In late 2014, the price of a barrel of oil fell dramatically – by more than 40 per cent. Although global demand for oil has fallen, oil companies are still producing the same amount, still spending millions and millions of dollars searching for new sources of oil.

We have all become addicted to oil. In 2011, we, the citizens of Earth, collectively used about 89 million barrels of oil every day. The most oil-hungry country in the world is Singapore. Every 1000 Singaporeans use 202 barrels of oil every day. Afghanistan is one of the countries that uses the least oil – a fraction of a barrel every day per 1000 people. In Australia, we use 44 barrels per day per 1000 people.

PLASTIC FANTASTIC

During the twentieth century, the petrochemical industry discovered that many other products could be made from crude oil besides fuels. The main one is plastics. All the plastic items we use each day – soft drink bottles, toys, food storage containers, pens, plastic bags – are made from crude oil. Many other non-plastic products are also made from oil, such as bicycle tyres, nail polish and the soft capsules that medicines and vitamins come in.

Each year, the world produces around 300 million tonnes of plastic products. Fifty per cent of that is for disposable products.

Many familiar plastics, such as vinyl, polystyrene and PET (polyethylene terephthalate), were invented between the 1920s and 1950s. Companies patented their plastic creations and gave them brand names like Nylon, Teflon and Plexiglas. The manufacture of plastics produces carbon dioxide emissions. The durability of some plastics means that they take a long time to breakdown. Tonnes of plastic waste end up in waterways and in the oceans where it has a negative impact on plant and animal life. When plastics do eventually breakdown, methane is released.

In 1948, a plastic called acrylonitrile butadiene styrene (ABS) was invented. It is still used today in millions of products – including Lego blocks.

TURNAROUND

Eleuthère Irénée du Pont (1771–1834) escaped to America with his family during the French Revolution to avoid the guillotine. He opened a gunpowder factory in Delaware, which supplied millions of pounds of gunpowder to American armies. The company became known as DuPont. It went on to supply almost half the explosives used by Allied forces during World War I. They produced 70 per cent of US explosives used in World War II, and then went on to be involved in the manufacture of the first nuclear bomb. A DuPont chemist invented the world's first synthetic fibre – nylon. After World War II the company was in the business of producing plastics, nuclear bombs and synthetic materials used in space flight. They were part of the team that put lead in petrol and were big producers of CFCs. In 2005, DuPont was acknowledged for reducing its greenhouse gas emissions by more than 65 per cent from the 1990 levels. At the same time, it used 7 per cent less energy and produced 30 per cent more product. However, the company is still criticised for the amount of toxic waste it produces.

AIRPOCALYPSE

But at least we fixed the smog … didn't we? Not exactly.

Up until the 1960's, China's economy was almost entirely agricultural. The Chinese "Industrial Revolution" didn't begin until 100 years after the English one. With its massive population, China has industrialised ten times faster than England. It has overtaken the United States as the world's biggest industrial producer. Private car ownership in China has risen from 0 in the 1960s to 120 million today. All this has led to shocking pollution and soaring greenhouse gas emissions.

In February 2014, thick brown smog lay over the capital city of Beijing for more than a week. It is a regular occurrence. The pollution comes from industrial areas surrounding the city, the 200 coal-fired power plants that produce electricity for the region and the city's 5 million vehicles.

The World Health Organization calculates the level of pollution by measuring how many micrograms of "fine particulate matter" there are in the air. The recommended level is 25. During this period, the levels in Beijing were consistently over 450 and reached a peak of 505.

The fine particulate matter in smog is called PM2.5. It consists of particles of soot and droplets of chemicals that are 2.5 micrometres or less in diameter – small enough to enter the human bloodstream. (One millimetre = 1000 micrometres.)

Beijing is not the most polluted city in the world, not even the worst in China. In 2013, there were 12

Chinese industrial cities with higher pollution levels. India is also rapidly industrialising. In late 2013, Delhi won the dubious honour of being the world's most polluted city, with levels of pollutants 60 times higher than recommended levels.

ABC

For several months every year, a phenomenon known as Atmospheric Brown Cloud (ABC) now hangs over large areas of China, India, South-East Asia and parts of Africa. First recorded in the late 1990s by a team led by Veerabhadran Ramanathan, ABC is a layer of air pollution containing soot, dust and gases such as nitrogen oxides and sulphur dioxide. The pollutants are produced by the widespread use of inefficient cooking methods involving burning fossil fuels (often kerosene), wood and cow dung. Other sources of pollutants are forest fires intentionally lit for land clearing, and industries in countries with no regulations to control emissions.

Atmospheric Brown Cloud can be up to three kilometres thick and can cover entire continents. Not only is ABC harmful to people's health, it also reduces the amount of sunlight by up to 25 per cent. This severely affects the growth of food plants and reduces the amount of rain that falls during the monsoon season.

CHANGE IN DIRECTION

Veerabhadran Ramanathan never found out the name of the African girl who asked him that question in New York, but that one student had a big impact on him. He made changes in his personal life, putting solar panels on his house and catching the bus to work. He also changed the direction of his climate research. He turned his attention from examining the science behind climate change to finding ways of using science to contribute to solutions that could drastically reduce emissions.

While governments argue about limiting carbon dioxide emissions, Ramanathan is concentrating on reducing other factors that contribute to global warming, such as black carbon and methane.

One of his initiatives is Project Surya, which focuses on reducing black carbon in the atmosphere by making small efficient stoves available to thousands of those women whose simple fires contribute to ABC.

THE GHOST OF CLIMATE YET TO COME

Scientists can now read the climate of the past. They can see how climate has changed over millions of years and understand what influenced those changes. They can also accurately measure the state of our atmosphere today. Using this information, they can create computer models to make predictions about the climate of the future.

But Earth's future climate will not just be the result

of the natural forces that changed climates in the past. It will be driven by the unknown factors of our ever-growing future emissions of greenhouse gases and other pollutants.

"Unfortunately, my generation has been somewhat careless in looking after our one and only planet. But, I am hopeful that is finally changing. And I am also hopeful that your generation will prove far better stewards of our environment."
Ban Ki-moon,
UN Secretary General,
address to international students, 2007

12

RUNNING ON EMPTY

21 January 2007
Samsø, Denmark

That's our wind turbine. Our very own. I love to just stand and watch it turn, whirling around like a toy for a giant. My dad didn't like the idea at first. He said that Tranbergs had raised dairy cows and grown pumpkins here on Samsø for three generations and he didn't see any reason to change. Even when Søren Hermansen's plan to make Samsø run 100 per cent on renewable energy won that government competition, he was still unimpressed.

Søren was my sister's teacher at the high school. His dream was that Samsø would stop using the electricity that comes to the island through cables from the mainland.

"What did we win in the competition?" my dad asked Søren.

Søren had to admit there was no money.

"So the prize is nothing more than a pat on the back and a good-luck wish?" Dad said. "How do we make electricity with that?"

The plan was that us Samsings, the 4000 people who live on the island, would buy the wind turbines and the solar panels ourselves. Dad thought it was a crazy idea. He wouldn't give Søren any money.

Søren didn't give in. He talked the government into giving him some money – just enough for him to give up his job and start working on the project. No one else on the island thought it would ever work, but he kept talking about it. He talked at schools and churches, at council meetings. He stopped people in the street or sat down next to them in cafes. He must have spoken to every single Samsing.

"We will make money from it," he said.

Dad didn't believe him.

I'd seen the wind turbines at Tunø Knob. I believed Søren. I gave him the 46 kroner I had saved from my pocket money.

Some people wanted to be involved, and Søren raised enough money to buy the first three wind turbines. Dad still wouldn't listen to him.

Then one day, Søren asked if I wanted to go up inside one of the wind turbines. Of course I did! But I couldn't go up without an adult to accompany me. I begged Dad to come with me and eventually he gave in.

It was amazing. We had to climb up the 140 rungs of a ladder, attaching and unattaching carabiners as we went. At the top there was hardly room for the three of us because of all the machinery, which was vibrating like crazy. Then Søren pressed a button and the ceiling opened. We were 50 metres up, and the wind was so strong it felt like it was going to rip my hair out by the roots. The turbine blades are HUGE close up and they were whirling around so fast. It was scary and exciting.

I could tell Dad was impressed. He had to shout to make himself heard over the roar of the wind.

"One thing we certainly have got a lot of here on Samsø is wind."

He was starting to change his mind.

"So once you've paid for the turbine, then all that electricity you make is free. Is that right, Dad?" I knew it was. "And the power that comes along the cables from the mainland, that's always going to cost money, isn't it?"

If there's one thing my dad likes, it's getting something for nothing.

"I'm no greenie," he said, "but I can see the logic in making our own electricity."

"You don't have to be a greenie to know that it's better to use energy that's free," Søren said.

Now we've got a wind turbine of our very own. At least, it's on our land. Dad put money into it, and so did other Samsings. The island now has 11 turbines on land and 10 offshore. We make more electricity than we use on the island and we sell the excess to the power companies.

A windy day's a good day here on Samsø. Dad and I like to go out and watch our wind turbine turning, making us money.

Torben Tranberg, student, aged 11

The Earth's stores of fossil fuels are not limitless. Once we have used up all the reserves, it would take hundreds of millions of years to create more. So some people around the world, like the people of Samsø, have taken charge of making their own electricity. But there are many people who are still using fossil fuels as if they will last forever, with no regard to the escalating levels of carbon dioxide and other greenhouse gases in our atmosphere.

SHRINKING SOURCES

The fossil fuel with the largest reserves is coal. That could last until around 2100. We might have enough natural gas to last until 2070. Oil is the fossil fuel that is in shortest supply. Estimates are that there is enough oil to last for about 50 years. We could use up all the oil by 2065. That might sound like a long way off, but about 70 per cent of people living today will still be alive then. They could well see the end of oil.

The oil companies say there is no need to worry, these figures only refer to the reserves of oil in the ground that we know about. They are confident there is plenty more still to be discovered. But this means getting at smaller deposits of oil and gas that are harder to find and more difficult to get out of the ground. These are sometimes referred to as "unconventional fossil fuels".

As it becomes more difficult to find and extract the oil, it also becomes a more expensive process. Oil companies will have to keep increasing the price of their products to compensate. Despite the volatility of the price of crude oil, these companies are still determined to keep producing it. The more expensive the products of crude oil become, the more people will try to use less. Sooner or later, the extraction of the Earth's remaining oil will become uneconomical.

UNCONVENTIONAL

In Australia, the fossil fuel that companies are searching for now is gas – pockets of methane trapped in coal seams deep underground. This is called coal seam gas. To extract it, the companies first drill down between 300 and 1000 metres. They then often drill horizontally in different directions to get at more gas. To help the gas escape, the coal seam is fractured by pumping in a mixture of water, sand and chemicals at high pressure. This process is known as fracking. Shale gas is found in shale, a soft rock containing a high level of carbon from ancient plants. It is extracted in a similar way to coal seam gas.

Other unconventional fossil fuels that are currently being extracted include tar sands oil and tight oil. Tar sands are a mixture of sand, clay and a very thick oil called bitumen. They are mined in Canada and Venezuela, often from open pits. The bitumen is extracted and then refined into oil. Tight oil is oil trapped in shale. Fracking is used to release this oil.

It takes two to four tonnes of tar sands to produce one barrel of oil.

The most environmentally damaging unconventional fossil fuel is called (confusingly) oil shale. This oily rock is rich in kerogen, and it is burned to liquefy the kerogen. The use of oil from oil shale involves burning the fossil fuel twice, once to extract it and then again when it is used. This results in enormous levels of greenhouse gas emissions.

Aerial view of coal seam gas wells near Tara, Queensland.

Peak Downs coal mine in Queensland. It is the sixth-largest coal mine in the world.

TIGHT, EXPENSIVE, POLLUTING

The process of extracting unconventional fossil fuels requires large amounts of energy, often millions of gallons of water and many millions of dollars. The processes can contaminate waterways and groundwater with methane and the chemicals used in the fracking process, which then find their way into drinking water supplies. Though companies don't have to make public the chemicals they use, it is known that many of them are toxic.

During the collection and transportation of unconventional gas, some can leak into the atmosphere. Shale oil and coal seam gas wells usually only produce for a short period of time. Extractors then move on to another piece of land. Large areas of farming land are lost to these mines.

And of course, all of the unconventional fossil fuels produce greenhouse gases when they are burned to create energy for our use.

RENEWABLE ENERGY

Fortunately, fossil fuels aren't the only sources of energy that we can use. There are other sources that are ready to use right now. We don't have to wait millions of years for them to cook. We don't have to mine or drill for them.

Energy from the sun and wind are natural, non-polluting and, because they will never run out, they are known as renewable energy sources. Solar thermal power stations are large-scale electricity generators. Thousands of curved mirrors concentrate the heat from the sun, which is then stored by heating a column of fluid (usually molten salts – not the sort of salt we add to food, but mixtures of other salts, such as sodium and potassium nitrate). The stored heat is used to generate electricity 24 hours a day, even when the sun isn't shining.

Wind power is not a new technology. Sailing ships were powered by wind for thousands of years. Windmills have pumped water and ground grain since about 500 BCE.

Wind farms use the power of wind to turn large blades on turbines. This energy is used to generate electricity. Wind turbines are the modern version of windmills. Much smaller wind turbines can be used on houses, boats and caravans to generate electricity for direct use.

Other sources of renewable energy still under development are wave power, which converts the energy of ocean tides to electricity, and geothermal energy, which uses the heat deep in the Earth.

Renewable energy production results in no greenhouse gases being released.

Albany Wind Farm, in Western Australia.

THE SITUATION NOW

Around the world, greenhouse gas emissions are continuing to rise. The world is using twice as much electricity as it was in 1970. In one year, the inhabitants of planet Earth are together emitting more than 30 billion tonnes of greenhouse gases. From April to June 2014, the average amount of carbon dioxide in the atmosphere was above 400 ppm for the first time in human history. Remember, 350 ppm is believed to be the maximum "safe" level.

The largest contributor to global greenhouse emissions is energy – the electricity and gas that we use in our homes, schools and workplaces. Energy production contributes more than 60 per cent of the world's emissions. There is no sign that emissions are going to stop their steep increase.

In September 2014, just before a United Nations summit on climate change was held in New York, the heirs to oil king John D Rockefeller's oil fortune withdrew all their funds that were in fossil fuel investments, and vowed to reinvest only in sustainable businesses.

The international oil industry is spending billions of dollars searching for the Earth's dwindling supplies of oil in more difficult places, with diminishing chances of finding any. They are reluctant to diversify and start investing in renewable energy.

Around the world, governments are slowly changing to renewable energy sources. Though it still has a great many coal-fired power stations and huge pollution problems,

China is on the way to being the world leader in renewable energy, producing twice as much as the USA. Germany and India are also high on the list of countries that use wind power. But 87 per cent of the world's electricity is still generated by burning fossil fuels.

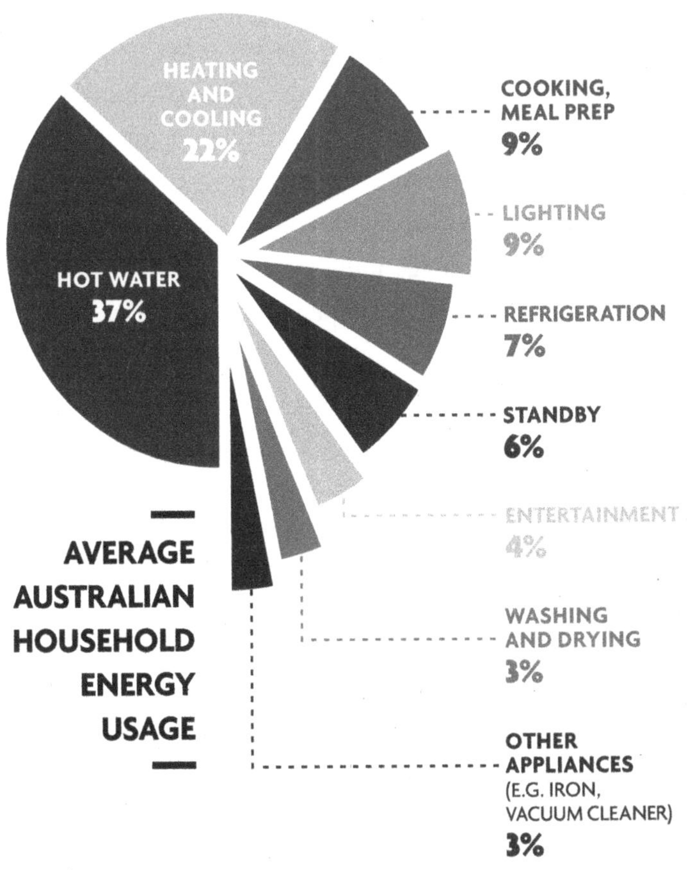

LAND OF SUN AND SURF

Australia is a very sunny country. We are not short of wind either. We are ideally placed for renewable energy production, and yet the amount of electricity we generate from renewable sources is small by world standards – just 15 per cent. Most of Australia's electricity is produced by old and inefficient coal-fired power stations. As well as emitting more carbon dioxide than any other source, they produce other harmful waste products that can escape into the atmosphere, such as nitrogen oxides, sulphur oxides and the trace metals arsenic and mercury.

Australia was one of the first countries to introduce a carbon tax to encourage industry to reduce its collective carbon footprint. The Renewable Energy Target (RET) scheme was created to ensure that at least 20 per cent of Australia's electricity is generated from renewable sources by 2020. Then, in 2014, Australia became the first and only country to abolish a carbon tax. The RET has been lowered, even though it has done what it was required to do – reduce greenhouse gas emissions by increasing use of renewable energy.

Because of our old and inefficient coal-fired power stations, Australia produces 60 per cent more emissions than the USA for each megawatt hour of electricity generated.

In Australia, we have more power stations than we need. The Australian Federal Government could begin the transition to renewable energy by closing the dirtiest coal-fired power stations and encouraging the development of renewable energy. But instead it is choosing to keep the old and polluting power stations open, while weakening its Renewable Energy Target and reducing the solar feed-in tariff that pays people for electricity generated on their rooftop solar panels that feeds into the national grid.

LAGGING BEHIND

Ninety-one per cent of Australia's electricity is generated by burning fossil fuels.

> *"Coal is good for humanity, coal is good for prosperity, coal is an essential part of our economic future, here in Australia, and right around the world."*
> **Tony Abbott, Australian Prime Minister, October 2014**

Plans to build Australia's first large-scale solar power plant were scrapped in 2014 because of lack of government support. There aren't a lot of wind farms either. South Australia has the highest level of renewable energy production including more than half of Australia's total wind power. It is on a par with European countries, producing almost 30 per cent of its power from renewable sources.

Australians who want to use electricity from renewable sources have had to take matters into their own hands.

Between 2006 and 2013, something amazing happened. The amount of electricity Australians purchased from electricity companies fell by seven per cent. Before that, electricity consumption had only ever risen. This is because people have been installing solar panels and also have been reducing their energy consumption by using electricity more carefully. More than 1.1 million Australian households have invested over $7 billion of their own money in solar panels. South Australia leads the way here too. A quarter of all dwellings in that state have solar panels installed.

In New Zealand, only 27 per cent of electricity is generated by burning fossil fuels. The rest is generated by renewable energy, mainly hydropower and geothermal power.

FORCINGS AND FEEDBACKS

The factors that influence climate change are many and complex. Some are beyond our control, such as the position of the Earth in its orbit, which results in fractionally more or less solar radiation arriving in our atmosphere; and occasional volcanic eruptions that shoot dust, soot and smoke into the atmosphere, which reflect solar energy back into space. But the factors caused by human activities, such as burning fossil fuels, are within our control. Scientists call these influences on climate – whether they are natural or man-made – "forcings" because they force a change in Earth's climate.

If a climate forcing causes warming, this can melt sea ice at the poles. If it causes cooling, it can cause an ice age.

James Croll was the first to describe what scientists today call a "climate feedback" nearly 150 years ago. His example of a climate feedback was how the whiteness of snow and ice reflects sunlight away from the Earth. Losing this heat makes it colder, and so more snow falls, more ice forms and … it gets even colder. This feedback loop could continue and lead to an ice age.

But the reverse can also happen. If the temperature rises and the ice melts, there is less white ice to reflect solar radiation away from the Earth, and more dark ocean to absorb it. This makes the Earth even warmer. This is one of the reasons why climate scientists are concerned about melting polar ice.

A MATTER OF URGENCY

The Earth's oceans are vast. Because of the slow circulation of water from the surface of the ocean to the depths, it can take a long time – decades or even centuries – for them to warm in response to climate feedbacks. Once these processes are set in motion, they are hard to stop. The result will be escalating global warming, which in Australia will result in more heatwaves and more destructive bushfires. A warmer atmosphere means more evaporation from the oceans. This will result in more water vapour in the atmosphere, which will lead to heavier rains and more flooding.

Global warming causes sea levels to rise. Melting polar ice contributes to this. Also, when water gets warmer each molecule expands a tiny amount. Because there is so much water in the oceans, this can result in a measurable rise in sea levels around the world.

Even if we were able to completely stop emitting greenhouse gases today, those gases we have already emitted will survive in the atmosphere for hundreds of years and global warming will continue for some time.

Earth's average temperature has risen 0.8°C in the last hundred years. Sea levels have risen 22 cm. Reducing our greenhouse gas emissions is a matter of urgency.

LETTER FROM THE POPE

In June 2015, after consulting with scientists and climate-policy experts, Pope Francis issued an encyclical (letter) on the environment. He addressed it not only to Catholics, but to "every person living on the planet". The 67-page letter discussed the issue of climate change, and asked what sort of world we want to leave for future generations. The pope challenged all people to change their consumer-driven lifestyles that result in much waste and pollution. He appealed to governments to urgently produce a common plan for a meaningful reduction in greenhouse-gas emissions.

PARIS, 2015

In December 2015, the UN Climate Change Conference took place in Paris, France. Participants at the previous Conference had been unable to agree on a plan to reduce global emissions, so the organisers in Paris were determined to produce an agreement that would have a real affect on global warming. After several nights of negotiations, the 196 participating countries agreed on the next steps to combat climate change. This is known as the Paris Agreement. The goal is to limit global temperature increase to 1.5°C. Previously, a 2°C rise was the target. To enable this to happen, countries must commit to achieving zero greenhouse-gas emissions by 2050–2100, and outline how they will reach this target. There will be progress reviews every five years, and countries will have to come up with better strategies if they are not meeting their goals.

PERMITS TO POLLUTE

The current Australian government has set a goal of 26–28 per cent reduction in greenhouse-gas emissions below 2005 levels by the year 2030. However, the government has no commitment to reduce reliance on coal-fired electricity generation, and is still giving out permits for new coalmines to be opened, and existing ones to expand. It is relying on buying carbon credits to meet its Paris targets. Carbon credits are permits that polluting countries can buy from other countries that have reduced their emissions below their targets. This will allow Australia to continue polluting the atmosphere. In his encyclical, the Pope makes his opinion of buying carbon credits clear. He says it is "a ploy which permits maintaining the excessive consumption of some countries".

EXCUSES, EXCUSES

There are many excuses that people make to avoid having to do something about climate change.

▸ Australia's emissions are insignificant

Some people say that Australia's emissions are small compared to the USA and China; therefore, there is no point in us reducing our carbon footprint. This is not true. Australia is fifteenth on the list of countries that produce the most emissions. Our country produced around 587 million tonnes of greenhouse gases in 2013. That's more than France, which has almost three times our population.

Together, Australians make a significant contribution to global emissions. As a prosperous country we should set a better example.

▸ My emissions are insignificant

You might think that you're just 1 person on a planet of 7 billion people and anything you do is insignificant. And anyway, it's the people who are polluting the most that should do something about it.

Per person, Australia has one of the highest rates of emissions in the world – 18.6 tonnes per year for each one of us. That's more than Americans, double what people in Germany or Japan produce, and three times the emissions of each person in China.

Australians, you and I, are among the most energy-wasteful people in the world. So if you think it's the people who are polluting the most who should cut their emissions, that's us. The ones who produce all that polluting coal? That's us too. We don't only use a lot of coal ourselves, every year we export more than 300 million tonnes of the stuff to other countries.

▶ Plenty of people think climate change doesn't exist

A recent survey concluded that 97 per cent of climate scientists believe that climate change is happening, and it is caused by human activities. That's a strong argument on its own. If doctors told you there was a 97 per cent chance you were going to get cancer, would you ignore their advice?

But just suppose it turns out that all those scientists have got it wrong and there is no risk to humans from climate change. Won't we have wasted a lot of time and money?

Even if we take the negative impacts of climate change out of the equation, replacing fossil fuel electricity generation with renewable energy will still result in positive effects:

- There will be less air pollution. That will mean less respiratory disease and other health impacts.
- In particular, people working in the energy sector and those living near electricity generators will be healthier.
- Our energy sources will be free and never run out.
- There is no toxic waste in the generation of renewable energy, so the world's drinking water will be safer.
- Building and maintaining renewable energy generation infrastructure will provide jobs.
- Billions of litres of water will be saved if we stop coal-fired electricity generation and fracking.

▸ Someone in the future will fix it

An argument often heard is that humans are inventive. They have invented so many incredible things in the past, some clever person in the future will think of a way to reduce emissions.

Governments in countries with coal reserves (like Australia) have been searching unsuccessfully for at least 25 years for "clean" coal-burning technology that will reduce emissions from coal-fired power generation by 30 per cent. They haven't found it yet. Other fields of research aim to find technologies that will allow us to continue to pollute by removing carbon dioxide from the atmosphere. Some schemes that have been seriously considered involve fertilising the oceans with tonnes and tonnes of iron filings. This would increase the amount of plankton, which would then consume more carbon dioxide as they photosynthesise. Pumping liquefied carbon dioxide deep underground (geosequestration) has also been suggested.

Another idea is to limit the solar radiation that reaches the Earth's surface. Proposals include injecting chemicals, such as hydrogen sulphide and sulphur dioxide, into the atmosphere, or placing banks of mirrors in space. Such schemes would cost billions of dollars without guarantee of success, and with plenty of possibilities for negative effects – some we are already aware of, others unforeseen.

Or we could just cut our emissions.

13

PEOPLE POWER

16 July 2015

Ringwood, Victoria, Australia

—

"You're only doing it because of that girl," my sister Becky says.

I'm hanging out my basketball uniform on the washing line. I take no notice of her.

It wasn't Sofia that got me started on reducing my carbon footprint. It was Vasily. He knows how competitive I am. He challenged me to do it. I've read some of the stuff that Sofia keeps sending via him every week. And I have to say, past generations of humans couldn't have done a better job of trashing the planet if they'd sat down and planned it. And there are plenty of businesses that are still trying to get away with wrecking the place. I accepted the challenge.

It was too hard to work out what my actual carbon footprint was, so I decided to work on my family's instead. The whole household. And also, it fits in with the science project I have to do. So, you know: two birds, one projectile.

I did buy my new runners. My feet are still growing, there's not much I can do about that. To make up for it, I bought some T-shirts and shorts from Savers and I donated my old runners to an organisation that distributes them to people in need.

I started by reducing the electricity we use, getting everyone to turn off lights and computers, etc. I figured out how to use that thing that we have that turns off the TV at the wall. Tiff says I've become a stand-by Nazi.

Mum works long hours and I didn't want to give her a hard time about drying the washing in the dryer, so I've come up with a plan. It's radical. I've traded all my chores for doing the washing. I don't stack the dishwasher any more, or take a turn with the vacuuming. I don't even have to put out the rubbish or walk the dog. All I do is the washing.

The first thing I had to do was cut down the amount of washing that we do. I mean, eight loads a week for four people, one who's only six? I made some rules.

1. You can wear some things more than once. (Not undies obviously or sportswear or things that you dribbled soy sauce down.)

2. You cannot have a clean towel every day. (This means you, Tiffany.)

3. There's a fine if you use more than two towels a week. (A dollar per towel. And I get to keep it.)

I've got it down to five loads. That saves water as well as electricity. Also, I found a washing powder that's made in Australia and has no phosphates in it. And we buy it in bulk (which means less packaging).

I've discovered there's a bit more to doing the laundry than throwing everything in the washing machine. First of all you have to wash dark things like jeans and black T-shirts (my clothes) separately from light things (girls' clothes and sheets). Otherwise the girls' undies end up grey and they don't like that. You have to get it all dry too. And not in the dryer obviously. Hanging out the washing is a real skill. You have to hang things by the corners, except for girls' fancy tops which you hang from the underarms so they don't stretch or something.

The first time I hung out a whole load of washing it rained all day. So now I check the weather forecast. I'd taken a pledge never to use the dryer. So what was I going to do when it rained? I found a thing to hang clothes on indoors in Grandma's shed. She says it's called a clothes horse. Don't ask me why. Mum and my sisters all complained about having the clothes horse in the sun room, so I ended up putting it in my room. I can hang sheets over the stair rail. Mum says it makes the place look like a laundry, but I told her it saves about $200 a year, and she's stopped complaining.

Mum cooks something vegetarian three times a week. Becky's organised a container for food scraps and she feeds the worms in our new worm farm. Tiffany is still complaining about being forced to take three-minute showers. I don't buy drinks in plastic bottles. The hardest thing for me has been cutting down on burgers.

The average Australian household produces more than 18 tonnes of greenhouse gases a year. I reckon we've halved ours.

Vincent Dwyer, student and clothes-drying expert, aged 16

Scientists have learned a great deal about the world's climate systems over the last 200 years. More than anything, they have learned that it is complex. Many factors, from the Earth's orbit to ocean circulation, from cow burps to making cement, affect our climate. It is easy to think that the situation is hopeless, that the climate is out of our control.

We have seen how the actions of ordinary people have helped turn the tide of inaction on environmental issues such as air pollution and ozone depletion. We can all reduce our own carbon footprint. If our governments, our big businesses and our electricity providers are unwilling to change their ways in order to reduce emissions, it is up to us, the citizens of Earth, to show them how it's done.

It is young people who will have to live with the effects of climate change. Even if you are under 18, there are still plenty of things you can do to help reduce greenhouse gas emissions.

BE INFORMED

The first thing we all need to do is to be informed. Make sure that you know the facts about climate change and the science behind it, so that you can argue the case when people use the excuses listed in the previous chapter. Watch out for climate change issues in the news.

Don't take my word for it. Do your own research, but be wary. Looking up climate change issues on the internet will bring up the opinions of climate sceptics and deniers as well as scientific evidence. Use trustworthy sites, such as the EPA, NASA and IPCC. If a website is one person's point of view, check if they have provided references for the source of their information. Does their list of references include scientific studies? Have the results been published in a respected scientific journal?

DON'T WASTE ELECTRICITY AND GAS

In the average Australian home, 20 per cent of electricity is used on appliances. Making sure your TV, computer, PlayStation, microwave oven, etc. are not left on stand-by is one way to reduce your carbon footprint. Six per cent of your household's electricity usage can happen when you aren't even using these things. That's a complete waste of a valuable resource. Dry the washing outside instead of in a dryer. As well as reducing greenhouse emissions, your electricity bill will also be smaller.

"In order to achieve that goal of stabilising the climate at two degrees or less, we simply have to leave about 80 per cent of the world's fossil fuel reserves in the ground. We cannot afford to burn them and still have a stable and safe climate."
Lesley Hughes, ecologist and Australian Climate Councillor, 2013

Only switch on the heater or the air conditioner when necessary. If it's cold, put on a jumper before you consider turning on the heating. Turn off lights in rooms that aren't being used. If your household can afford it, encourage fitting LED lighting and installing solar panels.

Reducing electricity usage cuts our emissions, saves us money and shows governments and industries that people are serious about stopping climate change.

USE THE CAR LESS

For most Australian households, transport is the area where they produce most greenhouse gases – an average of 34 per cent of their emissions. Every time you get into a car, you are contributing to greenhouse emissions. Think about each car trip you make. Could you walk or ride your bike? Could you catch public transport? Could you at least give someone else a lift?

There are now more than 1 billion cars on the road around the world.

REFUSE, RE-USE, RECYCLE

It isn't just the electricity you use at home and the amount of fuel your family car uses that contribute to your household's carbon footprint. You should also include the amount of energy it takes to make the products you use in your daily life. Manufacturing anything takes energy from somewhere, usually in the form of electricity usage. Then the goods have to be transported to where you buy them, by road, sea or air. That uses up more energy.

The most effective way to reduce emissions is not to buy stuff in the first place. We are

Australians use between 10 and 15 billion drink containers a year (glass, plastic, aluminium). Only 40 per cent of them are recycled.

Bottled water costs about 2000 times more than tap water.

learning to say no to plastic bags, but what else can we refuse? Don't buy items that have unnecessary packaging. Don't buy disposable items, such as bottled water. Fill up a reusable drink bottle from the tap. Buy rechargeable batteries and a charger.

If you have appliances, clothing, toys or games that you don't use any more, don't throw them away. Take them to a charity shop where they can be sold to people who have a use for these things. (This helps the charity too.) While you're there, see if there are any second-hand items that you can use rather than buying new things.

It takes much less energy to create things from recycled resources than to make them new. Take plastic bottles, for instance. The plastic itself has to be created before the bottle can be made, and plastics are all by-products of oil. Aluminium also has to be smelted from aluminium ore before it can be made into aluminium cans. That process uses huge amounts of energy.

Aluminium doesn't change in any way when it is melted down to be remade into another product. That means it can be recycled indefinitely. Making something from recycled aluminium uses only five per cent of the energy required to make something out of new aluminium smelted from ore. It also only produces five per cent of the greenhouse gases. About 75 per cent of the aluminium ever produced is still in use today, thanks to recycling.

Most municipalities and shires collect recyclables, usually plastics, aluminium cans and foil, tins, paper and cardboard. Find out what you can put in your recycling bin to ensure you are recycling everything possible. Look for the recyclable symbol on containers, etc. Your shire or council might also have a drop-off point where you can recycle other things that can't go in your recycling bin. My council depot accepts batteries, CDs and DVDs, scrap metal, printer cartridges and larger plastic items that won't fit in the recycling bin, such as toys.

In 2013, 279 million single-use batteries and 65 million rechargeable batteries were imported into Australia.

Bale of plastic bottles and liquid paperboard (LPB) cartons from household collections, ready for recycling.

DON'T WASTE FOOD

We waste a lot of food. Up to half of the contents of Australian household rubbish bins is food waste. This is not only a waste of food, it is a waste of the energy resources that went into growing, preparing and transporting it. When food scraps end up in a rubbish tip, they get buried, which means they decompose anaerobically (without the aid of oxygen). When that happens, methane is produced, and as we know, methane has a high global warming potential.

Composting is decomposition that takes place with the aid of oxygen, microorganisms and worms. No methane is produced. You can compost left over food, peelings, etc. in a compost bin. The compost you make is then added to your garden to provide nutrients. You can also use a worm farm. This turns food waste into a liquid fertiliser and worm castings (worm poo), which also adds nutrients to your garden.

TAKE RESPONSIBILITY FOR YOUR ENERGY USAGE

Other people in your household might do the washing and the cooking, so you might think you can't control those activities. You can take control of some of them. Offer to hang out your own clothes instead of putting them in the dryer. (You also have to keep your eye on the weather in case it rains, and remember to bring the clothes in again!)

If your household doesn't compost kitchen scraps,

perhaps you can take charge of that. Organise a container for food scraps, and empty it into the compost bin regularly. When the compost is ready, distribute it around the garden.

If you don't have a compost bin, make a compost pile. Ask for a worm farm for your birthday! If you don't have the space for composting, perhaps you can take the food scraps to a place that does compost, such as your school or a community garden. Some local councils have green waste collections that can include food waste.

GET POLITICAL

You might be too young to vote now, but in a few years you'll be old enough. Be prepared for when you can vote, and think carefully about who you want to vote for. You will be able to vote in three different levels of government – local, state and federal. Find out about the climate change and sustainability policies of the people standing for election in your area. Who is promising to introduce policies that will improve our atmosphere? Who is in favour of renewable energy? Who wants to improve public transport? Who opposes fracking?

Your opinion counts. Write letters to the newspapers. Email politicians and tell them your views on what the government should be doing about climate. Remind them that you will soon be able to vote. Sign and help distribute petitions on environmental issues. You don't have to be 18 to sign petitions.

Your generation might be the one that makes the

difference, the one that succeeds in saving Earth's climate where previous generations have failed.

SPREAD THE WORD

Your school community might already be doing great things to reduce its carbon footprint, but can it do more? All the energy-saving steps that apply to a household also apply to a school. What about your extended family members? Your sports club? Your faith group? Don't be afraid to point out where energy (and money) is being wasted.

BECOME A CITIZEN SCIENTIST

Remember the Marsham family who recorded the first signs of spring for more than 200 years? You can also become a citizen scientist. In Australia, ClimateWatch is an organisation that collects information about birds, animals and plant life. Thousands of people contribute observations of species they observe in their own gardens, in parks or on bushwalks. This information is sent to scientists who use it to see how climate change is altering the natural world.

You can take part in citizen science projects online. Zooniverse is a citizen science website involving people from all around the world. There are a number of projects that are collecting information to enable scientists to learn more about the Earth's environment and climate. You can listen to whales, measure plankton or transcribe weather observations from historic ship logs!

PARTICIPATE

Take part in the climate change movement so that you can keep up with new developments and help spread accurate information. You can join the Australian Youth Climate Council or volunteer for Conservation Volunteers Australia (under-17s will have to be accompanied by an adult). If you live in a rural area, there may be a branch of Lock the Gate Alliance that opposes fracking on farmland close by. Ask if you can help out in any way.

It there are no groups for you to join, start your own sustainability group at school.

YOU

Your name could be here. You could become someone who makes a real difference and helps to stop global warming.

THE CLIMATE PUZZLE SOLVED

"What's the use of having developed a science well enough to make predictions if, in the end, all we're willing to do is stand around and wait for it to come true?"
F Sherwood Rowland, Nobel laureate, 1986

The evidence is in. There's no doubt that climate change is happening. Two hundred years of climate research undertaken by thousands of scientists in dozens of scientific fields, backed up by 800,000 years of ice core evidence and millions of years of data from marine sediments is conclusive. The level of carbon dioxide in the atmosphere, measured on Mauna Loa, has been steadily climbing since 1958. In 2015, it passed 400 ppm. The last decade has been the hottest since meteorological records began in the 1800s. Climate scientists are still working, still learning, still collecting data, but the facts are undeniable. Like a jigsaw puzzle with just a few pieces left to put in place, the picture is perfectly clear. Global warming is real.

Ignorance, greed and indifference have caused, and continue to cause, greenhouse gases and other pollutants to damage our atmosphere. Our climate is already in an endangered condition. Around the world people are experiencing more extreme drought, bushfire and flooding; rising seas threaten coastal communities; heat-related illness increases; food crops and water supplies are under threat. Climate change will affect every person on the planet. It's up to us, every one of us, to take action to reduce greenhouse gas emissions.

The geological epoch that we are currently in began 11,700 years ago and is called the Holocene. Some scientists say it should be called the Anthropocene (from the Greek words *anthropos* meaning human being and *kainos* meaning recent). They argue that this is the period in which people have been responsible for altering the atmosphere, polluting rivers and oceans, and causing extinctions of plants and animals. The Anthropocene hasn't become an official epoch (geologists argue that there is no evidence in the geological record). There might not be evidence in the rocks, but there is in the ice and the ocean sediment.

If this epoch is to be named the Anthropocene after us, wouldn't it be great if instead of meaning the era when humans brought about catastrophic climate change, it became the era when we took responsibility, healed our precious atmosphere and made our climate safe?

We need to act fast. What we do now will determine the climate in the future. We must hold global warming to below 2°C above the level it was before the Industrial Revolution. To do that, we have to reduce the amount of greenhouse gas emissions for which we are responsible, individually and together.

Previous generations are responsible for the state of our climate, but it's young people who will inherit a planet changed by global warming. You have a job to do. Yes, you. You are the most important person in the campaign to fix our climate. You can make a difference.

TIMELINE

DATE	PERSON	NATIONALITY	EVENT
12,000 years ago			FIRST AGRICULTURE PRACTISED
1661	John Evelyn	English	Writes about polluted air in London. Claims it is unhealthy.
1760–1840			INDUSTRIAL REVOLUTION
1769	James Watt	Scottish	Applies for patent for new and improved steam engine
1824	Joseph Fourier	French	Calculates Earth would be a lot colder if it lacked an atmosphere
1837	Louis Agassiz	Swiss	Announces his theory of a Great Ice Age
1853			SMOKE NUISANCE ABATEMENT ACT
1859	John Tyndall	Irish	Discovers that some gases block infra-red radiation. Suggests increases in amount of these gases could cause climate change.
1859	E L Drake	American	First oil strike in USA
1875	James Croll	Scottish	Publishes astronomical theory of the ice ages. Says there was more than one.
1880	Thomas Edison	American	Patents his improved electric light globe
1880			FOG AND SMOKE COMMITTEE FORMED
1886	Karl Benz	German	Invents the first car
1896	Svante Arrhenius	Swedish	Publishes first calculation of global warming as a result of human emissions of CO_2. Predicts it will take 3000 years to double.
1908	George Bernard Reynolds	English	Oil is discovered in the Middle East
1913	Henri Buisson and Charles Fabry	French	Ozone layer is discovered
1913	Henry Ford	American	Begins mass production of cars
1921	Thomas Midgley Jr	American	Develops leaded petrol
1924	Milutin Milanković	Serbian	Publishes first version of his theory of solar radiation causing ice ages
1928	Thomas Midgley Jr	American	Synthesises the first CFC
1938	Guy Stewart Callendar	English	Proves CO_2 greenhouse global warming is happening
1943			BIG SMOG IN LOS ANGELES

1949	Arie Haagen-Smit	Dutch	Explains make-up of photochemical smog
1952			KILLER LONDON SMOG
1958	Charles David Keeling	American	Began accurately measuring CO_2 in the atmosphere
1959	Bert Bolin	Swedish	Makes prediction that CO_2 will increase by 25–30% on pre-industrial levels by the end of the century
1966	Cold Regions Research and Engineering Laboratory	American	First deep ice core drilled
1970s			REMOVAL OF LEAD FROM PETROL BEGINS
1973	F Sherwood Rowland and Mario Molina	American/Mexican	Discovers CFCs breaking down ozone in stratosphere
1975	Veerabhadran Ramanathan	Indian	Discovers CFCs are also greenhouse gases
1976	John Imbrie et al.	American	Proves that Milanković's theory was correct
1980s			FIRST CLIMATE COMPUTER MODELLING
1980	US Windpower	American	Builds world's first wind farm
1982	Joseph Farman et al.	English	Discovers the ozone hole
1985	Veerabhadran Ramanathan	Indian	Calculates effects on climate change of greenhouse gases other than carbon dioxide
1988			IPCC FORMED
1989			MONTREAL PROTOCOL BECOMES EFFECTIVE
1997			KYOTO PROTOCOL ADOPTED
2001	William F Ruddiman	American	Hypothesises that humans have been affecting the atmosphere for 8,000 years
2004	EPICA	Collaboration of 10 European countries	Extracts ice core that reveals conditions 800,000 years ago
2006			MAXIMUM SIZE OF OZONE HOLE
2007	Abengoa Solar	Spanish	Builds world's first commercial solar thermal power plant
2007	IPCC and Al Gore	International/American	Jointly receive Nobel Peace Prize for efforts in spreading knowledge about man-made climate change
2010			END OF CFC MANUFACTURE WORLDWIDE
2014			HOTTEST YEAR ON RECORD

GLOSSARY

A

Anaerobic
A word describing an environment that has no oxygen.

Anthracite
Hard black coal, which has a high carbon content and burns with less smoke than softer coals such as brown coal. It produces less greenhouse gas.

Astronomy
The science of celestial objects (stars, planets, moons, asteroids, etc.), space and the universe.

Atmospheric pressure
The weight of air above a particular place on Earth. The highest atmospheric pressure is near the surface of the Earth where the weight of the air above compresses the air below. The further you travel up into the atmosphere, the lower the atmospheric pressure. Weather associated with low pressure is warm and wet. Weather associated with high pressure is cool and dry.

B

Bakhtiari
A tribe of traditionally nomadic people from south-west Persia.

Biosphere
The parts of the Earth where living things exist. This includes the Earth's crust, oceans, waterways and the air.

Brazier
A metal basket-like container for burning charcoal or wood, usually for use outdoors.

C

Carabiner
A metal clip used in mountaineering to join ropes together.

Carbon sink
Anything that absorbs more carbon than it releases. A carbon sink acts as a store of carbon, preventing it from entering the atmosphere. Examples of carbon sinks are forests and oceans.

Carbonic acid
Carbonic acid is formed when carbon dioxide is dissolved in water. In the nineteenth century, carbon dioxide was called carbonic acid.

Cataract
A condition of the eye where the lens becomes cloudy, resulting in poor vision.

Computer modelling
Using computer software to examine information from the past and the present to predict what might happen in the future.

Crude oil
Oil in its natural state as it comes out of the ground and before it is refined to remove impurities. In the USA, crude oil is called petroleum.

Cyclone
A storm formed above a tropical ocean involving winds blowing in a circular motion.

D

Depletion
When something is used up or decreased.

Developed world
The countries in the world that have advanced industry and technology, and are usually economically stable.

Drawers
An old-fashioned word for large and loose underpants.

E

Energy
Energy is needed to make things work. In discussions about reducing greenhouse gases, energy usually refers to the things we use to power our vehicles, appliances and factories, e.g. electricity, gas and fuels such as petrol and diesel. Renewable energy comes from sources that will not run out, such as the Sun, wind and water. Non-renewable energy comes from sources that are finite, that is, they will eventually

run out. Coal, gas and oil products such as petrol are non-renewable energy sources.

Eon

In the geological timescale, an eon is the largest division of time, lasting billions of years. An eon is divided into eras, which are measured in hundreds of millions of years. Eras are divided into periods, measured in tens of millions of years. Periods are divided into epochs, measured in millions of years.

Epoch

A geological period of time or a period marked by a distinctive character.

F

Fossil fuels

Non-renewable sources of energy that were formed under the Earth from the remains of plants and animals over millions of years. Coal, oil, gas and products made from them such as petrol and diesel are fossil fuels.

G

Geophysics

The physics of the Earth and its atmosphere, including the formation of the Earth and the natural phenomena that take place on Earth, such as earthquakes, volcanoes and climate.

H

Haberdasher

A person selling small items used for sewing and decorating, such as buttons, thread and ribbons.

Hunter-gatherers

A group of people who survive by hunting animals and collecting wild plants for food.

Hydrocarbons

Compounds that contain only carbon and hydrogen atoms. Examples are natural gas, oil and many plastics.

Hypothesis

A suggestion for an explanation of why something happens, based on a carefully thought-out assumption that will need to be investigated, or a more confident proposal based on facts established by experimentation or collection of information.

I

Internal combustion engine

An engine in which the fuel that produces the energy is burned inside the engine. Car engines are internal combustion engines. Steam engines burn the fuel outside the engine.

Iron Age

A period in history when people had learned to smelt iron from iron ore and used it to make tools and weapons. In Britain, the Iron Age was between 800 BCE and the 1st century CE.

Isotopes

Two or more forms of a chemical element that have a different number of neutrons in their nuclei.

M

Middle East

An area extending from the east of the Mediterranean Sea into Asia. It includes Egypt, Turkey, Iraq, Iran, Saudi Arabia, Israel and Palestine.

N

Nitrogen cycle

Nitrogen is the most abundant element in Earth's atmosphere, and is essential to all living things, but it is not available directly from the air. Nitrogen must combine with other elements, such as oxygen, hydrogen and carbon, before it can be absorbed by plants and animals. The process of nitrogen circulating through the atmosphere, soil, plants and animals by forming and reforming different compounds is called the nitrogen cycle.

O

Omnibus

An old-fashioned word for a bus. Originally omnibuses didn't have engines, they were pulled by horses.

P

Paleontology

The study of fossil plants and animals.

Persia
The former name of the country now called Iran.

Photosynthesis
A process in plants where carbon dioxide and water are absorbed and converted into carbohydrates using the energy from sunlight.

Phytoplankton
Microscopic plants that live in fresh water or in oceans, often in large amounts. Even though they are so small, they still grow using photosynthesis.

R

Respiration
The act of breathing, which involves taking in oxygen from the air so that it can combine with carbohydrates to produce energy. Waste products of this process are carbon dioxide and water vapour, which are breathed out.

S

Smelt
To heat ore at very high temperatures until it melts in order to obtain metals such as iron, aluminium and silver.

Symbiotic
Animals and/or plants that have developed in a way that makes them rely on each other for survival.

T

Tellurium
A rare white element that often occurs when combined with gold or silver. In industry it is used to make alloys of iron, copper and lead.

Think tank
A name sometimes given to a research organisation where experts give information and advice to governments or businesses.

Tuber
A fleshy part of a plant, which is a root or stem growing underground. Some tubers are edible, e.g. potatoes.

U

United Nations (UN)

A group of nations formed in 1945, after World War II, to promote peace and cooperation around the world. The UN currently has 193 member nations.

W

Worsted

A type of woollen yarn often used for making suits and coats.

WEBSITES

CITIZEN SCIENCE

www.climatewatch.org.au
www.zooniverse.org

CLIMATE FACTS

For today's carbon dioxide level:
As measured at Mauna Loa Observatory in Hawaii
keelingcurve.ucsd.edu

As measured at Cape Grim, Tasmania
www.csiro.au/greenhouse-gases/

COMPOSTING

www.foodknowhow.org.au/waste-know-how/composting/
www.cleanup.org.au/au/LivingGreener/composting.html
yourenergysavings.gov.au/waste/reducing-recycling/kitchen-food-waste/composting

CSIRO STATE OF THE CLIMATE 2014

www.csiro.au/Outcomes/Climate/Understanding/State-of-the-Climate-2014.aspx

HOUSEHOLD GREENHOUSE GAS CALCULATOR

Calculate your household's carbon footprint:
www.epa.vic.gov.au/agc/home.html
www.wwf.org.au/our_work/people_and_the_environment/human_footprint/footprint_calculator/

PARTICIPATING

www.aycc.org.au
www.conservationvolunteers.com.au

RECYCLING

www.recyclingnearyou.com.au

WORM FARMING

yourenergysavings.gov.au/waste/reducing-recycling/kitchen-food-waste/start-worm-farm

SOURCES

EPIGRAPH

Archbishop Tutu quoted in "The power of religion and prayer to head off climate disaster", *The Guardian*, 18 September 2014, www.theguardian.com/sustainable-business/2014/sep/17/religion-prayer-climate-change-desmond-tutu

CHAPTER 2

Flannery, Tim 2005, *The Weathermakers*, Text Publishing, p. 70

CHAPTER 3

Shen Kuo, *Dream Pool Essays*, chpt. 24, quoted in Needham, Joseph 1959, *Science and Civilisation in China*, vol. 3, Cambridge University Press, p. 609

Aristotle inline quoted in Glacken, Clarence J 1967, *Traces on the Rhodian Shore: Nature and Culture in Western Thought from Ancient Times to the End of the Eighteenth Century*, University of California Press, pp. 129–30

Lucius Junius Moderatus Columella, *De Rustica*, book 1, Preface 1:4, p. 29, Loeb Classical Library, www.loebclassics.com/view/columella-agriculture/1941/pb_LCL361.29.xml

William Gladstone quoted in "Mr Gladstone at Home", *The Times*, 6 Aug 1877, p. 8

Agassiz, Louis 1840, *Etudes sur les Glaciers*, quoted on NASA Earth Observatory website earthobservatory.nasa.gov/Features/Paleoclimatology/paleoclimatology_intro.php

CHAPTER 4

Fourier, Joseph 1827, *General Remarks on the Temperature of the Terrestrial Globe and the Planetary Spaces*, translation in *American Journal of Science*, vol. XXXII, no. 1, p. 16

Arrhenius, Svante 1908, *Worlds in the Making: The Evolution of the Universe*, Harper & Brothers, p. 54. Internet Archive archive.org/details/worldsinmakinge01arrhgoog

CHAPTER 5

William P Rend quoted in Morag-Levine, Noga 2003, *Chasing the Wind: Regulating Air Pollution in the Common Law State*, Princeton University Press, p. 110

Queen Victoria quoted in Strong, Roy 1998, *The Story of Britain: A People's History*, Random House, p. 394

Farey Jr, John 1827, *A Treatise on the Steam Engine: Historical, Practical and Descriptive*, Longman, Rees, Orme, Brown and Green, p. 8. Internet Archive archive.org/details/treatiseonsteame01fareuoft

CHAPTER 6

Harrison, Frederic 1886, "A Few Words about the Nineteenth Century", *The Choice of Books and Other Literary Pieces*, Macmillan, p. 438. Internet Archive archive.org/details/choicebooksando00harrgoog

Evelyn, John 1661, *Fumifugium, or the Inconvenience of the Aer and Smoake of London Dissipated*, p. 5, facsimile published by The Rota, University of Exeter, 1976 Internet Archive archive.org/details/fumifugiumorinc00evelgoog

Russell, Rollo 1880, *London Fogs*, Edward Stanford, at Victorian London website www.victorianlondon.org/weather/londonfogs.htm

Octavia Hill: letter to her mother, 24 May 1880, in ed. Maurice, Edmund 1913, *Life of Octavia Hill as Told in her Letters*, Macmillan, p. 437. Internet Archive archive.org/details/lifeofoctaviahil00hilluoft

Report: "The Smoke Abatement Exhibition", *Nature*, vol. 25, no. 636, p. 219

CHAPTER 7

Han dynasty quote: Needham, Joseph 1959, *Science and Civilisation in China*, vol. 3, Cambridge University Press, p. 609

Frank Howard quoted in Kovarik, William 2005, "Ethyl-leaded Gasoline: How a Classic Occupational Disease Became an International Public Health Disaster", *International Journal of Environmental Health*, vol. 11, no. 4, p. 388

CHAPTER 8

New York Times quote inline: *The New York Times*, 28 April 1959, in Somerville, Richard C J 2009, *Bert Bolin and his 1959 forecast of atmospheric carbon dioxide and climate change* at ams.confex.com/ams/89annual/techprogram/paper_143689.htm

Wolff, Eric: in interview 4 August 2010 on Open Knowledge website knowledge.allianz.com/environment/climate_change/?618/ice-cores-window-into-climate-history

CHAPTER 9

McNeill, J R 2000, *Something New Under the Sun: An Environmental History of the Twentieth Century World*, W. W. Norton, p. 111

Farman, Joseph 1999, transcript of *My Century: The Story of the Twentieth Century — Told by Those Who Made It*, BBC World Service, broadcast 6 July 1999 www.bbc.co.uk/worldservice/people/features/mycentury/transcript/wk27d2.shtml

CHAPTER 10

Hansen, J et al. 1981, "Climate Impact of Increasing Atmospheric Carbon Dioxide", *Science* 28 Aug 1981, vol. 213, no. 4511, p. 966

CHAPTER 11

Veerabhadran Ramanathan: personal email communication, 4 Dec 2014

Inline quote re: devastating impact: *Australian climate change policy: A chronology* at Parliament of Australia website www.aph.gov.au/About_Parliament/Parliamentary_Departments/Parliamentary_Library/pubs/rp/rp1314/ClimateChangeTimeline

AYCC: *About AYCC*, AYCC website http://www.aycc.org.au/about_aycc accessed 19 Feb 2015

McKenzie, Amanda, "Inaction will cost us our future", *The Drum*, 29 Sep 2010, ABC website www.abc.net.au/news/2010-08-20/35958

Ban, Ki-moon: 1 March 2007, *Address to the United Nations International School–United Nations Conference on "Global Warming: Confronting the Crisis"*, UN website www.un.org/apps/news/infocus/sgspeeches/search_full.asp?statID=70

CHAPTER 12

Tony Abbott quoted in *The Sydney Morning Herald*, 13 Oct 2014 www.smh.com.au/federal-politics/political-news/coal-is-good-for-humanity-says-tony-abbott-at-mine-opening-20141013-115bgs.html

CHAPTER 13

Lesley Hughes quoted in "Climate Commission report says 80 per cent of fossil fuel reserves must stay in the ground", 26 June 2013, ABC News website www.abc.net.au/news/2013-06-17/fossil-fuel-reserves-must-stay-in-ground-report/4757448

THE CLIMATE PUZZLE SOLVED

F Sherwood Rowland quoted in "In the Face of Doubt", *The New Yorker*, 9 June 1986, p. 70

IMAGE CREDITS

COVER

Cloud in sky © Maksym Darakchi/Shutterstock.com

CHAPTER 1

p.16 Earth © NPeter/Shutterstock.com
p.19 Diagram of greenhouse effect © daulon/Shutterstock.com

CHAPTER 2

p.34 Forest in the Carboniferous Period from *Meyers Konversations-Lexikon*, 4th edition, Book 15. Image sourced from Wikimedia Commons.

CHAPTER 3

pp.49-50 Deforestation images © iStockphoto.com/luoman
p.57 Boulders perched on mountain peaks © Reid, Harry Fielding. 1902. No Glacier: From the Glacier Photograph Collection. Boulder, Colorado USA: National Snow and Ice Data Center/World Data Center for Glaciology.

CHAPTER 4

p.63 John Tyndall's setup for measuring radiant heat absorption from Tyndall, John, *Contributions to Molecular Physics in the Domain of Radiant Heat*, 1872. Image sourced from Wikimedia Commons.

CHAPTER 5

p.86 Children working in textile factory © Wellcome Library, London, under a Creative Commons 4.0 license.

CHAPTER 6

p.92 Ware factory illustration © Wellcome Library, London, under a Creative Commons 4.0 license.
p.96 London smog from *The Illustrated London News* © Wellcome Library, London, under a Creative Commons 4.0 license.

CHAPTER 8

p.133 Ice core with visible volcanic ash layer © Heidi A Roop, University of New Hampshire/US National Science Foundation

p.137 Graph of correlation between temperature and CO_2 levels in Antarctica by ML Design. From *The Complete Ice Age* edited by Brian Fagan © Thames & Hudson Ltd., London

CHAPTER 10

p.166 Cape Grim GAW Global Station © World Meteorological Organization, under a Creative Commons 4.0 license.

CHAPTER 11

p.179 Graph of Keeling Curve © Scripps Institution of Oceanography, UC San Diego

p.181 Solar electrical energy generation © Stocksy.com/MaaHoo

p.184 AYCC PowerShift march © Patrick McCarthy

p.184 AYCC "flash mob" © AYCC

CHAPTER 12

p.199 Coal seam gas wells near Tara, Queensland (Aerial view) © Moira McDade, Lock the Gate Alliance, under a Creative Commons 2.0 license.

p.200 Queensland: Peak Downs mine with machine © Lock the Gate Alliance, under a Creative Commons 2.0 license.

p.203 Albany Wind Farm in Western Australia © David Steele/ Shutterstock.com

CHAPTER 13

p.224 Bale of containers ready for recycling © Carole Wilkinson

ABOUT THE AUTHOR

p.254 Carole Wilkinson in rally © John Wilkinson

INDEX

AUTHOR'S NOTE

I'm writing this as I travel through the currently rather turbulent troposphere on a bumpy flight to Adelaide.

I have had some experience of climate change. I lived in a household heated by a coal fire for the first 12 years of my life. I experienced the 1962 London smog that spread up to the Midlands and was so thick it stopped all traffic, so that I had to walk three miles home from school unable to see more than a few inches in front of me. Stopping my use of spray cans in the late 1970s was perhaps my first environmental action.

I also have some experience with science. I studied science at school. Before I was a writer I was a laboratory technician for about 15 years. My name appears on a published scientific paper ("Routine histology by Carole Ross"). I operated a slide projector at an Australian & New Zealand Association for the Advancement of Science (ANZAAS) conference in 1969.

I worked with scientists. I saw at first hand the communication and collaboration between scientists working in the same field. I trust scientists.

I was very pleased when Black Dog Books publisher Maryann Ballantyne asked me if I would like to write a book about climate change. I didn't know everything about climate, far from it, so as with every nonfiction book I've written, the research process began. It was huge. I didn't realise I'd taken on a topic that stretched back millennia, but was still making the news almost every day as I wrote.

After months of reading my way through millions of years of climate change on Earth, my trust in the climate scientists has been confirmed. So my thanks to every one of them for their vision, dedication and rigour.

I would also like to thank all the people responsible for getting scientific papers online, whoever they are. I must have saved days, maybe weeks, by not having to traipse up and down aisles of printed journals and then photocopy them. Many dollars were saved … and trees. (I should also thank that unnamed person who let me use their university library card, but I better not.)

There are always little things that can't be learned from academic papers and I am grateful to those experts who were kind enough to answer my emails and who were generous with their knowledge. Thanks to Bruce A Bauer, NOAA; Brian Fagan, University of California; Ralph Keeling, Scripps Institution of Oceanography; Veerabhadran Ramanathan,

Scripps Institution of Oceanography; William F Ruddiman, University of Virginia; Tim Sparks, Coventry University; and Malcolm W Wallace, University of Melbourne. Sincere thanks to the expert readers who took the time to read drafts – Dr Malcolm Brandon, Dr Mark Curran, Amanda McKenzie, Mark Norman and David Spratt.

A large part of writing this book involved sitting in a room by myself, but when I needed them, a crowd of fabulous people appeared to assist. Special thanks to my publisher and editor, Maryann Ballantyne, who travelled with me on this journey and helped me keep on track when I wandered too far into the labyrinth of dire things humans have done to our atmosphere. I am grateful to the readers of early drafts, John Wilkinson and Andrew Kelly, for their suggestions. Thanks to the Walker Books team who have improved my words, added fabulous style and images and given me encouragement: Virginia Grant, Nicola Santilli and Amy Daoud.

Thanks as always to John and Lili for their unflagging support and encouragement throughout the writing process.

Carole Wilkinson taking part in Walk Against Warming, Melbourne 2009.

ABOUT THE AUTHOR

CAROLE WILKINSON is an internationally award-winning and bestselling author. Carole writes both fiction and nonfiction and her stories are loved by young people all over the world.

Carole embarked on her writing career at the age of 40 and she has been making up for lost time ever since. Carole is interested in the history of everything. She is also a member of a climate action group, campaigning for sustainable living and a safe climate. These two passions have been brought together in *Atmospheric: The Burning Story of Climate Change*.

OTHER BOOKS BY CAROLE WILKINSON

The Dragon Companion
Ramose: Prince in Exile
Ramose and the Tomb Robbers
Ramose: Sting of the Scorpion
Ramose: Wrath of Ra

THE DRAGONKEEPER SERIES

Dragonkeeper
Garden of the Purple Dragon
Dragon Moon
Dragon Dawn (prequel)
Blood Brothers
Shadow Sister

YOUNG ADULT

Sugar Sugar
Stagefright

PICTURE BOOK

The Night We Made the Flag

TRUE TALES SERIES

Ned Kelly's Jerilderie Letter

THE DRUM SERIES

Black Snake
The Games
Alexander the Great
Fromelles: Australia's Bloodiest Day at War

THE BEAT SERIES

Hatshepsut: The Lost Pharaoh of Egypt

Find out about Carole's books on her website
www.carolewilkinson.com.au